WHEN THE SAHARA WAS GREEN

WHEN THE SAHARA WAS GREEN

HOW OUR GREATEST DESERT CAME TO BE

MARTIN WILLIAMS

PRINCETON UNIVERSITY PRESS

PRINCETON & OXFORD

Published by Princeton University Press
41 William Street, Princeton, New Jersey 08540
99 Banbury Road, Oxford OX2 6JX

press.princeton.edu

All Rights Reserved
First paperback printing, 2023
Paper ISBN 9780691253930
Cloth ISBN 9780691201627
ISBN (e-book) 9780691228891

British Library Cataloging-in-Publication Data is available

Editorial: Ingrid Gnerlich, Marcia Garcia, and Whitney Rauenhorst
Production Editorial: Ali Parrington
Text and Jacket/Cover Design: Karl Spurzem
Production: Danielle Amatucci
Publicity: Sara Henning-Stout & Kate Farquhar-Thomson
Copyeditor: Karen Verde

This book has been composed in Arno

Printed in the United States of America

In memory of Françoise Gasse, Théodore Monod, and Pascal Lluch:
You shared my love and respect for our greatest desert. Valete!

CONTENTS

ILLUSTRATIONS

Maps

Figures

Tables

Plates

PROLOGUE

An Isolated Mountain

Just after dawn in late March 1970, two men were walking slowly alongside the ancient shoreline (plate P.1) of a lake that had dried out roughly 8000 years ago, when something unusual sticking out from the edge of the dry lakebed caught their eye. The former lake was bounded by high rocky ridges that were part of a rather unique type of mountain that geologists call a ring-complex. From the air, these types of mountains look like gigantic dartboards because they consist of relatively circular concentric ridges and valleys (fig. P.1). This isolated mountain in the heart of the Sahara Desert was known to the Tuareg camel herders in this region as Adrar Bous (plate P.2). It was surrounded by rolling sand dunes and undulating sandplains and, lacking water and being mostly devoid of vegetation, was uninhabited.

One of the men, Professor John Desmond Clark, was a distinguished archaeologist at the University of California, Berkeley. The younger man was the author, who had been a professional soil surveyor mapping soils along the lower Blue and White Nile valleys in Sudan. The previous night had been very windy, blowing highly abrasive sand grains across the surface of the now dry lake. The lake sediments consisted of easily eroded fine silts. What the night's sandstorm had revealed was the rib cage of a long dead hippopotamus. Even more exciting was what was embedded in the rib cage of the hippo: a barbed bone harpoon point. The two men looked at each other. They agreed: 'This one stays'. Desmond and I had just completed three months of archaeological excavations and geological mapping and sampling, and this was our last day at Adrar Bous before the long journey back to California for Desmond and his archaeological team and to Sydney, Australia for me.

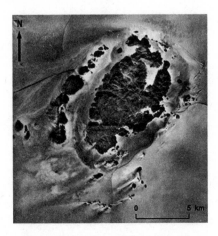

FIGURE P.1. Adrar Bous seen from the air. Photo
mosaic compiled by the author from air photos.
IGN, National Photo Library.

This chance discovery reminded us of another lucky find three
months earlier, on our very first morning at Adrar Bous, when we were
the two first members of our team to reach the mountain. Once again,
it was an early morning walk, as we began to familiarise ourselves with
what was to be our home for the next twelve weeks. Just as the sun
rose, a tiny piece of what looked like white bone caught the light. The
bone was embedded in a very hard, dark grey clay. The bone was in
fact a horn-core. Later painstaking excavation with dental picks revealed the entire skeleton of what turned out to be the oldest complete domestic cow ever recovered from the Sahara. It proved to be
about 5000 years old.

It was entirely by accident that we were to spend three dry and windy
months at the isolated mountain of Adrar Bous. If you draw an imagi-
nary circle of 1500 kilometres (or nearly 1000 miles) radius, with Adrar
Bous at the centre, the circle rim will only intersect the Mediterranean
coast of North Africa and the Atlantic coast of West Africa. At Adrar
Bous, you are about as far inland as it is possible to be in North Africa,
which is one reason why it is so dry. The summer monsoon rains seldom

reach this far north; the westerly winter rains seldom reach this far south. Even if they did, by then they would have already lost most of their initial moisture supply so that not much rain would fall. As a general rule, the further inland you go, the drier it becomes, an effect known to climatologists as 'continentality'. So, why did our team choose Adrar Bous as its destination?

Revolution and a Change of Plans

Our initial plan had been to continue our earlier work in Libya in the southeastern Libyan Desert. In the course of two summer expeditions there in 1962 and 1963 led by Captain David N. Hall, a young British army officer with the Royal Engineers, we had made some exciting discoveries. Our first summer was spent at an isolated mountain in the far southeast of the desert known as Jebel Arkenu (maps 1 and 3). Here we had found many beautiful rock paintings and abundant evidence of prehistoric human occupation. The following summer we mapped and named two huge sandstone plateaux in the far south of the Libyan Desert. Once again, we noticed abundant signs that this now arid region had been occupied on several occasions by prehistoric people, from Early Stone Age times onwards, a lapse of time amounting to nearly a million years. As well as prehistoric stone tools, we saw numerous rock engravings and multi-coloured paintings of people and herds of domestic cattle which offer a glimpse into a way of life that is no longer possible in this now desolate and arid region.

In order to build on our previous work and desert experience, David had assembled a very capable team of people with a range of practical skills; he had invited archaeologist Desmond Clark to bring a small team of graduate students and included me to assist with studying the soils and the landforms. Most of our group were young army and navy officers, all of whom had recently completed a degree in engineering at my old University of Cambridge. We had our team, vehicles, and supplies ready to go when disaster struck. Colonel Muammar Gaddafi staged a coup, the Libyan Royal Family were forced to flee, and foreigners were no longer welcome.

By great good luck Desmond had met the French Saharan archaeologist, Professor Raymond Mauny, at a lecture in California. Mauny had suggested that Adrar Bous in the Ténéré Desert (map 1) of the Republic of Niger (map 2) might be worth a visit, so that became our destination. A French expedition sponsored by the Berliet truck company had already spent some time there and had reported abundant surface finds of prehistoric stones and bones. However, said Mauny: 'There is no stratigraphy!' By this he meant that all the material was on the surface, having been concentrated by the combined work of wind and water, so that nothing was in its original position. From a scientific perspective, this meant that little of archaeological value could be culled from the surface finds. Fortunately, Desmond was not put off by this pessimistic verdict, and our later excavations were to prove it wrong. There was plenty of good, undisturbed material, provided you were willing to dig for it, which we did, for many weeks.

We drove down from Tunisia (map 2) in North Africa through Algeria (map 2) to the small town of Agadès (map 1) at the southern tip of the Aïr Mountains in Niger (map 1). We then drove north through the mountains to the oasis of Iferouane (map 1) in the heart of the ranges. From there we drove east across some appallingly difficult terrain. Our overloaded Land Rovers suffered badly. When another half-shaft on one of our three Land Rovers had snapped under the strain of driving through this rocky land, David made the wise decision to return the vehicles to Agadès for repairs in our temporary workshop there, while the archaeological team proceeded on foot or by camel across the desert to Adrar Bous, a journey of about three days, allowing for frequent stops to inspect any features of interest.

Journey by Camel to the Mountain

The journey by camel to Adrar Bous was revealing. Desmond and I were the first to set forth, led by Zewi bin Weni (plate P.3), our Tuareg guide and a former *goumier* or scout in the French colonial army. After leaving the main mountain range behind us, we proceeded north along a wide, flat, dry valley flanked to the west by the mountain front and to the east by a high sand dune (plate P.4) which ran for tens of miles northwards, parallel to the dry valley. Two dark, terrace-like features were visible along the flanks of the dune (plate P.5). Scattered across the surface of the lower terrace were broken bits of pottery and stone tools which Desmond recognised as being Neolithic in age, that is, at least 5000 years old. The Neolithic people were farmers and herders, so we immediately began to ask ourselves how they had managed to live in this now harsh, dry landscape.

The upper terrace was equally intriguing. Desmond believed the stone tools on its surface to be 'Epi-Palaeolithic', that is, dating to the final stages of the Late Stone Age, between about 12,000 and 8000 years ago, just before the transition from a lifestyle based on hunting, fishing, and gathering wild plants for food and medicine to one based on plant and animal domestication. From my perspective as an earth scientist, these terraces were in fact the abandoned floodplains of once flowing rivers. The sediments in the upper terrace were very fine-grained silts and clays, and were disposed in thin horizontal layers, quite unlike the coarse sands and gravels lining the floor of the dry valley. The environment must have been very different and the climate wet enough to allow a river to flow through what is now a desert.

The following days brought more surprises. On the second morning, we spotted the sand-blasted base of a pot emerging from the sand. The pot acted as a cover for another pot, which Desmond exhumed with his brush (fig. P.2). Inside the lower pot were hundreds of long-dried-up fruits of the *Celtis integrifolia* tree—a tree that today grows in areas that receive about 450–500 millimetres of rain a year. Also embedded in the sand were the fossil shells of the large land snail *Limicolaria flammata*, which today is found mostly in the tall grass savanna regions of central Sudan, where the annual rainfall is at least 450 millimetres.

A picture was now beginning to emerge in our minds, indicating that the climate had indeed been wetter during Neolithic and earlier times, prompting us to ask ourselves when and why it had become dry. As we got closer and closer to Adrar Bous, the scatter of stone tools and broken and occasionally well-preserved pottery on the surface of the present sand-plain became increasingly abundant. We began to detect grindstones,

FIGURE P.2. Neolithic pot excavated by Professor Desmond
Clark in the Ténéré Desert between Adrar Bous and the
Aïr Mountains.

arrowheads with hollow bases or with tangs for attaching them to the
former arrow shafts, polished stone axe heads, and many more stone
tools. Most of the smaller implements were delicately fashioned from a
beautiful green stone (later identified as a 'silicified vitric tuff') which
was obviously highly prized by the Neolithic and earlier stone tool makers. Although we later searched far and wide, we never located the
source of this rock at the time, but later discovered that it came from a
valley located 80 kilometres away, in the heart of the Aïr Mountains.

Our curiosity aroused by what we had seen on the camel trek to the
mountain, we were by now eager to discover what other prehistoric
finds lay in wait for our archaeological trowel and brush or, in my case,
what the ancient soils and sediments would reveal to my pick and
shovel.

The stark contrast between present aridity and the overwhelming
evidence of a recently wetter past has led me on a lifelong journey into
the past. I wanted to know why the Sahara had once been a green and
well-watered land able to support large animals such as elephants, giraffes, hippos, and crocodiles that are now found many hundreds of
miles further south in the savanna regions of East Africa. I was also curious to learn what might have caused the once fertile Sahara to dry out

and become an arid wilderness. Could humans have been a cause? Or had the climate changed? If so, why had it changed? Would it change again in the future? My attempts to find persuasive answers to these questions and others relating to the impacts of droughts on human societies provided the impetus that led me to write this book. The journey has been one of wonder and excitement for me; I hope it is for you, the reader, as well.

ACKNOWLEDGEMENTS

I owe a great debt of gratitude to Dick Grove, my former lecturer at the University of Cambridge, for inviting me to join a British Army expedition to Jebel Arkenu in the desert of southeast Libya in the northern summer of 1962. Captain (later Lieutenant-Colonel) David Hall of the Royal Engineers led this and two later Saharan expeditions (1963 and 1970) with quiet efficiency and great good humour. We have remained close friends ever since. The three months we spent at Adrar Bous in 1970, an isolated mountain in the Ténéré Desert of Niger, was another highlight of my time in the Sahara. Working closely with Professor Desmond Clark and his team of archaeology graduate students from the University of California, Berkeley, Andy Smith and Allen Pastron, I learned a great deal about African prehistory. Between us, we uncovered a record of prehistoric occupation and environmental change in the central Sahara extending back at least half a million years. We also recovered the oldest complete domesticated Saharan Neolithic cow.

Mohamed Tahir Ben Azzouz of the University of Constantine, Algeria, inducted me into the complexities of the saltbush landscapes of the Algerian Aurès Mountains that he was studying for his doctorate degree in 1969. In Tunisia, I had the privilege of working with a talented team of French earth scientists, including Professor Pierre Rognon, Professor Jean-Charles Fontes, Dr Françoise Gasse, Dr Geneviève Coudé, Dr Alain Lévy, Dr Jean Riser, and my good friend, Jean-Louis Ballais. I learned a great deal from all of them—and not just about this northern edge of the Sahara.

Professor Mike Talbot was a sterling companion during our camel trip to Wadi Azaouak during the terrible 1974 drought in Niger. It was on this trip that I was to discover how local botanical knowledge can help purify green and muddy water.

In 1976, thanks to the support and advice from Gerald Wickens (Kew) and Martin Adams (Hunting Technical Services), I was able to visit Jebel Marra volcano in arid western Sudan and to work on piecing together the environmental history of that now inaccessible region, with Dave Parry (Hunting Technical Services), Don Adamson (Macquarie University, Sydney), and Bill Morton and Frances Dakin from the Geology Department, Addis Ababa University.

During a visit to the eastern Sahara in 1987, I was fortunate to work with Professor Fred Wendorf from the Southern Methodist University, Texas, and Professor Romuald Schild from the Polish Academy of Sciences, Warsaw, at the unforgettable prehistoric sites of Bir Sahara and Bir Tarfawi in the hyper-arid Western Desert of Egypt. Before embarking on the fieldwork, I benefitted from the sage advice of Dr Bahay Issawi in Cairo. The succession of Last Interglacial lakes that existed here some 125,000 years ago revealed once again how the Saharan climate had oscillated between wet and dry.

Neil Munro introduced me to the desert country west of the Nile in northern Sudan in 2005. In the dunes west of the lower White Nile, Donatella Usai and her husband Sandro Salvatori showed me some of the complexities of Mesolithic and Neolithic occupation in this region in 2011, as did other members of their team, including Lara Maritan and Andrea Zerboni. I thank them all for their warmth and welcome. In 2012, Professor Matthieu Honegger and his team of Swiss archaeologists helped me to recreate in my mind the prehistoric landscapes of the desert country of northern Sudan.

At the other end of the Sahara, during a walking conference in the desert of western Mauritania in 2004, I met up with friends old and new: Professor Suzanne Leroy, Dr Françoise Gasse, Professor Mohamed Ben Azzouz, and Dr Hélène Jousse, all of us under the expert guidance of Pascal Lluch and Damien Parisse. Pascal was always pleased to share his vast practical knowledge of Saharan natural history.

Dr Matt Lloyd at Cambridge University Press, New York, very graciously put me in touch with Ingrid Gnerlich at Princeton University

Press. Ingrid quickly provided three knowledgeable, constructive, and anonymous reviewers for my draft proposal and final draft and has always responded to my enquiries with helpful and practical advice. Among the other Princeton University Press staff, my special thanks go to illustration specialist Dimitri Karetnikov, senior production editor Ali Parrington, editorial assistant María García, copyeditor Karen Verde and senior copywriter Theresa Liu.

With characteristic generosity, Professor Claudio Vita-Finzi in London provided me with a selection of his photographs from the Messak region in the Libyan Sahara and the Tassili in the Algerian Sahara showing prehistoric rock paintings and engravings. After a chance meeting at an international conference in Dublin in July 2019, Professor Olaf Bubenzer, University of Heidelberg, very kindly put me in touch with Professor Rudolph Kuper, University of Cologne, who provided me with a photograph of a prehistoric rock painting from the Tassili in Algeria showing women riding oxen.

My sincere thanks go to Professor David Thomas (University of Oxford), Professor Nick Drake (King's College, London), and Professor Nick Lancaster (Desert Research Institute, Reno, Nevada), all of whom provided welcome and generous advice on matters in which their expertise far exceeds mine, and to Mr Neil Munro (Addis Ababa, Ethiopia) who very kindly guided me towards some stunning satellite images.

My wife, Frances, provided rigorously honest criticism of each draft chapter, for which I am deeply grateful. In addition, Frances drew a number of the figures with her customary skill and elegance. A thousand thanks!

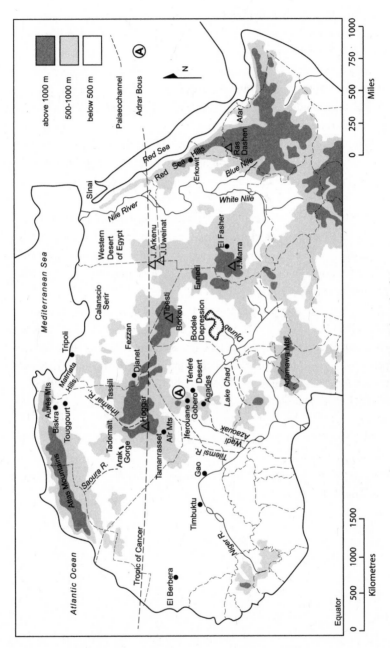

MAP 1. Major Saharan localities.

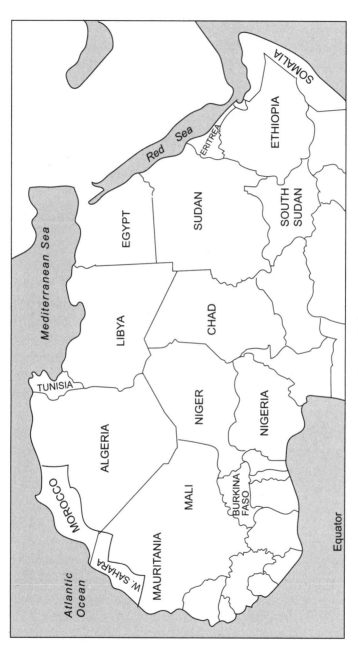

MAP 2. Saharan and adjacent countries.

WHEN THE SAHARA WAS GREEN

Introduction

The Sahara is vast. From the Atlantic Ocean to the Red Sea (map 1), it extends across a distance of 5000 kilometres. Its southern limit is at roughly 16° N and its northern limit at about 30° N, equivalent to a distance of about 3000 kilometres. It covers an area of roughly 9.2 million square kilometres, or almost four times that of the Mediterranean, which has an area of 2.5 million square kilometres. The Sahara is so big that any generalisation about it can be unwise. However, a glimpse at a rainfall map of North Africa (fig. I.1) shows that if we ignore the Saharan uplands, the lines of equal rainfall (*isohyets*) run parallel to one another and show a rapid decrease in rainfall with distance from the northern and southern coasts. The vegetation zones of North Africa (fig. I.2) also run parallel to the northern and southern coasts and, except in the uplands, which have their own microclimate and upland ecosystems, become increasingly adapted to aridity with distance inland. At present, the summer and winter rains do not penetrate very far into the Sahara, so that much of that vast desert is largely devoid of vegetation except in sheltered uplands and sporadic desert oases where groundwater comes to the surface. Between about 15,000 and 5000 years ago the situation was very different. The tropics received more radiation from the sun, the summer monsoon was accordingly stronger, and both summer and winter rains reached as far as the present-day arid heart of the Sahara. As a result, the vegetation belts also moved further inland, so that Mediterranean winter rainfall plants colonised the northern Sahara and tropical summer rainfall plants colonised the southern and central Sahara. In effect, the vegetation zones that we see today had at that time shifted more than a thousand kilometres further inland along both northern and southern margins of the Sahara—the time when the Sahara was green.

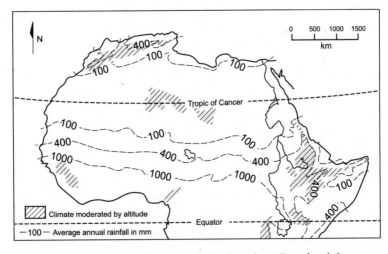

FIGURE I.1. Present-day rainfall zones in North Africa, after Williams (1988), fig. 3.4, adapted from *Atlas of Africa* (1973), p. 35.

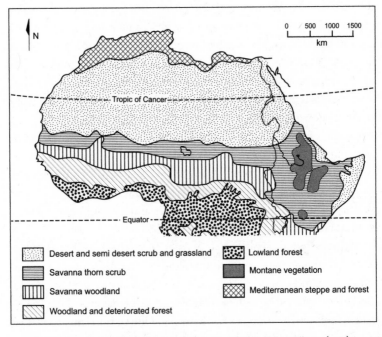

FIGURE I.2. Present-day vegetation zones in North Africa, after Williams (1988), fig. 3.5, adapted from *Atlas of Africa* (1973), p. 39.

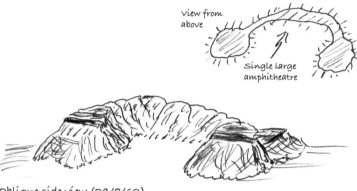

Oblique side view (29/8/62)
N of Kufra, SE Libyan Desert

FIGURE I.3. Sketch showing how prolonged erosion of an initially flat sandstone mesa in the Libyan Desert near Kufra has changed it into a single large amphitheatre (29 August 1962).

Dissected sandstone plateau, SE Libyan Desert (31/8/62)

FIGURE I.4. Sketch of a dissected sandstone plateau in the SE Libyan Desert (31 August 1962). The low sandstone mesas were originally part of a single continuous plateau.

My first visit to the Sahara in the northern summer of 1962 to the southeast Libyan Desert gave me the impression of being on another planet. Not a blade of grass for hundreds of miles. Stark black hills (figures I.3–I.7) rising from endless plains with a thin surface layer of fine gravel. Great humpbacked sand dunes aligned in seemingly endless rows parallel to the wind. Sand, dust, and wind; wind, dust, and sand. And yet it had not always been so. Hard to believe, but in all suitable

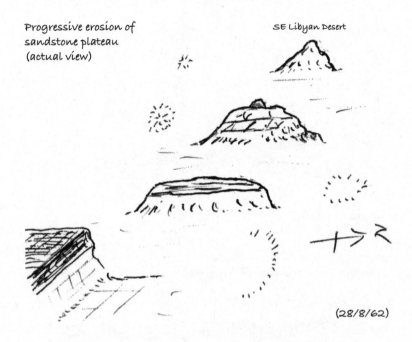

Progressive erosion of
sandstone plateau
(actual view)

SE Libyan Desert

(28/8/62)

FIGURE 1.5. Sketch showing stages in the conversion of a flat-topped sandstone mesa
to a conical sandstone butte, SE Libyan Desert (28 August 1962).

rock exposures, prehistoric artists had engraved or painted scenes of cattle camps and herds of giraffes and elephants. They left behind their stone arrowheads, grindstones and polished axe heads, ostrich eggshell beads, and the bones of fish, turtles, even crocodiles and hippos. How was this possible? The answer lay partly buried beneath the ever-shifting sands in the form of now dry lakes and defunct river channels. Their former presence prompted the questions: when, why, and how had the present desert once been able to support such an abundance of life? My attempt to answer these questions provides the reason for this book.

The book is intended for the nonspecialist reader interested in the natural world. My aim is to reveal how the Sahara Desert came into being and to show that, on a number of occasions in the past, it had been 'a green and pleasant land' well able to sustain an abundance of plant

Gara et Tuila, SE Libyan Desert (30/8/62)

Undercut sandstone isolated mounds, SE Libyan Desert (31/8/62)

FIGURE I.6. Sketches of a sandstone hill near Kufra in the SE Libyan Desert showing undercutting of the softer beds of isolated sandstone remnants (30 and 31 August 1962).

South of Agedabia, Libyan Desert (18/8/62)

FIGURE I.7. Sketch showing relief inversion of ancient dunes in the northern Libyan Desert (18 August 1962). The resistant beds acting as caprocks on the top of the dunes today were initially deposited in the hollows between the former dunes and were cemented by calcium carbonate in the groundwater. Later wind erosion removed the former dunes and resulted in progressive inversion of relief.

and animal life and to attract diverse groups of prehistoric hunters and herders until it became too dry to support much life. Questions raised and answered in this book include why the Sahara was previously much wetter, why it became dry, and whether it will become wetter once more. A related question is whether human activities might have caused the Sahara to become a desert. I also consider the impact upon prehistoric and modern human societies of extreme climatic events such as prolonged droughts.

The book is divided into three parts and concludes with a short epilogue. Part One provides a concise account of how the Sahara came into being and explains when and how the Saharan highlands and lowlands were fashioned, culminating in a description of the time when it was last a land of lakes and rivers and was aptly called the Green Sahara. Part Two looks at how the Sahara became progressively drier, with sand dunes developing from the alluvial sands brought down from the uplands by desert rivers. This was a time of constant tug-of-war between flowing water and wind-blown sand. Part Three looks at the Sahara today and considers how extreme climatic events such as prolonged droughts affect human societies and how human activities can aggravate (or minimise) the impact of such extreme events. The epilogue asks whether the Sahara could become green once more and what humans can do to live in harmony with our greatest desert as well as with the drier regions of the earth more generally.

PART ONE

THE GREEN SAHARA

The popular view of the Sahara is one of endless rolling sand dunes alternating with vast sandplains from which emerge occasional rocky hills (plate A.1). Nothing could be further from the truth. Only about a fifth of the Sahara actually consists of desert dunes and sandplains. The remaining four-fifths consist of rugged mountain ranges (plate A.2), vast sandstone or limestone plateaux, and extensive gravel plains (plate A.3) made up of fine wind-blown and alluvial sediments capped with a thin surface layer of fine gravel and stones. The word Sahara comes from the Arabic word '*sahra*', meaning a wilderness or desolate land to be traversed as quickly as possible—one that certainly did not invite one to linger there.

One of the most striking aspects of the Saharan landscape is the abrupt change from steep hill slopes to very gentle foot slopes, which is quite unlike the landscape of rolling hills we know so well from temperate Europe, North America, and Asia, with their deep soils and continuous cover of trees and grasses. Another very characteristic feature of the Sahara today is the presence of very old landforms located cheek by jowl next to very young landforms. By very old I mean many millions of years and by very young I mean a few thousand years or less.

The prehistoric human inhabitants of the Sahara made use of the resources provided by the entire country through which they roamed, including the plants and animals that had long adapted to these diverse landscapes and the rocks that provided the raw materials for their stone tools.

The next three chapters provide a sketch of how the Sahara first came into being. Chapter 1 sets the scene with an account of how the major elements in the Saharan landscape were fashioned. Chapter 2 reveals how dinosaurs once roamed freely across the Sahara, how the sea sometimes covered much of what is now the Sahara, and how it was later covered in tall trees that have since been fossilised. Chapter 3 describes the last time the Sahara could rightly be called a 'green and pleasant land'.

CHAPTER 1

Origins

The human intellect cannot grasp the full range of causes that lie behind any phenomena. But the need to discover causes is deeply ingrained in the spirit of man.

LEO TOLSTOY, *WAR AND PEACE*

My Introduction to the Sahara

My first introduction to the Sahara was in the northern summer of 1962. Soon after I had graduated from the University of Cambridge with a degree in geography, my mother received a surprise telephone call from Dick Grove, one of my former university lecturers, asking if I would like to join a British Army expedition to southeast Libya (map 2). I was away caving in the limestone dales of Derbyshire at the time, but my slightly delayed reply was an unqualified 'Yes'. I asked Dick what I was expected to do. His reply was commendably brief: 'Everything'. We set forth in early August. A young British Army officer in the Royal Engineers, Captain (later Lieutenant-Colonel) David Hall, led the trip. David and I hit it off very well and remain good friends to this day. Looking back, it is easy to see that the middle of summer was not the best time to go gallivanting off into the Sahara. We navigated using dead reckoning and the sun compass devised by Brigadier Ralph Bagnold for use in World War II. At night a theodolite star fix confirmed our location. There was little scope for error. Aiming for a few straggly palm trees on a bearing of 172

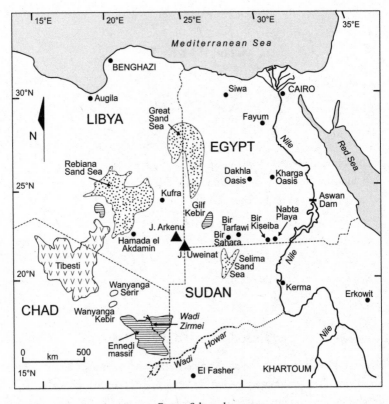

MAP 3. Eastern Sahara place names.

degrees for several hundred kilometres required skill and a calm but alert temperament. Find the trees, and you can dig a few feet down to find fresh water. Fail to find the trees and you would not last long in the Saharan summer without water. As we proceeded south from Augila to Kufra (map 3), I recorded shade temperatures at noon on four consecutive days of 40°C (105°F), 40°C (105°F), 44°C (111°F), and 47°C (116°F). It grew hotter as we drove further south to our ultimate destination, Jebel Arkenu (map 1), a small, isolated mountain which it was my job to map.[1]

I had never been in a desert before. I had seen photographs of desert dunes and had read about them, but this was a desert baptism by total immersion. My previous research had involved mapping evidence of glaciation in the wetter parts of southwest Ireland, where it had rained constantly, and I never saw the sun. Perhaps that is what predisposed me in favour of later working in those parts of the world where one could count on remaining reasonably dry. In any event, the Libyan Desert aroused my curiosity. Everywhere I looked there were signs that running water had once been active. In addition, sheltered wadis within the mountains were home to prehistoric paintings and rock engravings of people and their herds of cattle. There were even remains of fish bones in now hyper-arid areas. How was this possible? When was the last time that the Sahara had been able to support such copious life? Why could it no longer do so? When did it become dry? Why? I never imagined, when I first embarked on this journey into the heart of the desert, that I would still be seeking answers to these same questions a half century on. We do have answers, many of them solidly buttressed by independent lines of evidence rigorously dated by a variety of independent methods. But there remain even more unanswered questions to which we still seek convincing replies.

I noticed something else while collecting samples of rocks from Jebel Arkenu to help me construct a geological map of the mountain. Many of the rocks were highly decayed and very friable, not the sort of thing you expect to find in a hyper-arid desert. Furthermore, the central granite dome was surrounded by concentric layers of rock that had become detached from the parent rock and on a gigantic scale resembled layers peeled back from an onion. Some of these sheets were up to five metres thick and separated from the rock of the granite dome by gaps up to sixty centimetres in width. This type of detachment is sometimes called onion weathering, but this expression is misleading. The detachment of concentric layers of rock is caused by the release of pressure once a certain thickness of overlying rock has been eroded away. The cause is pressure release, not weathering, and the result is a series of dome-shaped pinnacles of rock. However, the cause of the rock decay is certainly linked to weathering, in this case deep chemical weathering under a very wet tropical climatic regime at a time when the entire region was covered in tropical rain forest and the underlying rocks were chemically altered to depths of many score metres.

As for the famous desert winds, about the only sign of their impact, apart from the Great Sand Sea and its rolling dunes located 250 kilometres (150 miles) to the northwest, were the low alcoves eroded by wind abrasion on the windward face of many small sandstone knolls (plate 1.1). Even here, the decayed nature of the rock next to the surface made the erosive work of the wind much easier. I never saw much evidence of wind abrasion where the rocks were hard and fresh. Minor undercutting by wind in weathered rocks or deep grooving of the soft sediments on the floors of former lakes was about all I could find in support of the much-vaunted erosion by desert winds. And so, almost by default, I became convinced that the major elements of the desert landscape that I was seeing for the first time had been carved by long-vanished rivers that had been actively eroding the desert mountains and plateaux for many millions of years before aridity finally set in. Whether this erosion was continuous or intermittent I was not then in a position to say. Today we can determine with some confidence that erosion by fast-flowing and highly energetic rivers of what ultimately became the Sahara Desert was at first prolonged and perennial and then became more and more sporadic as progressive desiccation extended its influence across the desert.

The Sahara is a wonderful natural museum. Thanks to its very aridity, delicate prehistoric rock paintings and very ancient fossil trees, some in growth position, are beautifully preserved at widely scattered localities across the desert. In addition, the geological evidence of how the Sahara came into being is very accessible at or close to the surface and not hidden beneath a thick mantle of soil and vegetation. The emergence of the Saharan landscape was a gradual process and took place in fits and starts. It began a long time ago, well before the first appearance of complex life-forms about 540 million years ago. In this chapter we provide a series of snapshots showing how the Sahara has evolved through geologic time. Later chapters provide more detail.

Saharan Landscape Diversity

Although some parts of the Sahara are very flat and exceedingly monotonous, the overall landscapes of the Sahara are remarkably diverse and there are always surprises in store. I will illustrate this with a few examples.

The Algerian oasis of Tamanrasset (map 1) nestles at the foot of the mighty Hoggar Mountains (map 1) which rise to a height of 2,908 metres above sea level and occupy an area of more than half a million square kilometres (map 1). One of the most striking features of the Hoggar are the isolated volcanic plugs scattered around the main massif like petrified sentinels from another era (plate 1.2). High up on the flat summit of a peak known as Assekrem is the small and very solid stone hermitage (plate 1.3) built in 1910 by Père Charles de Foucauld (1858–1916). He had served as a cavalry officer in the French colonial army in Algeria and later became a Catholic priest and hermit, much respected by the local Tuareg people. He was killed in 1916 by a group of dissident Senussi tribesmen from across the Libyan border. When I visited the hermitage at Assekrem in early January 1970, I found there a copy of Professor Pierre Rognon's impressive study of the evolution of the Hoggar.[2] I imagine that he must have left it there for the benefit of visitors with a broad interest in the Hoggar. I was to meet Pierre later at a conference in Addis Ababa, Ethiopia in December 1971, and in 1978–79 I spent a year in his department in Paris.

Located about 330 kilometres north of Tamanrasset are the Arak gorges (map 1). These desert canyons were cut deeply into the horizontal sandstone of the southern Tademaït Plateau (map 1) by a former large and fast-flowing river. The cliffs on either side of the gorge are impressive and rise up to five hundred metres above the canyon floor. There is no flowing water in the Arak gorges today, but prehistoric stone tools scattered along the dry sandy floor of the gorges point to wetter conditions in the past.

There is a remarkable oasis in the desert of Mauritania (map 2) in the western Sahara, known as the Oasis of El Berbera (map 1; plate 1.4). It is hidden away beneath the cliffs of a vast sandstone plateau covered in rock fragments coated in the dark patina of iron and manganese oxides known as desert varnish. Standing on the surface of this huge stony tableland, you have no idea that only a short distance away and about fifty metres beneath the surface of the plateau is a deep pool of fresh flowing water surrounded by palms and other green and luxuriant vegetation. A steep ramp of sand allows easy access down to the pool. Underground water flowing out of the sandstone cliff maintains this oasis as a perennial source of fresh water. The surrounding cliffs are undercut, and a series of shallow alcoves contain accumulations of recently precipitated calcium carbonate forming stalactites and stalagmites—surprising in this sandstone region. The groundwater must be tapping some source of limestone on its underground journey. What is most noteworthy about this hidden oasis is that unless someone told you it was there, it would be all too easy to walk on, entirely unaware that cool fresh water was close at hand.

Back to the Basement

Much of the Sahara is covered in very gently sloping sedimentary rocks mostly composed of sandstone or limestone, with minor patches of finer grained softer rocks which tend to erode fairly rapidly and so form broad, shallow depressions. Some of the oldest sedimentary rocks have been subject to intense heat and pressure from the weight of overlying rocks and have been altered—metamorphosed—to very resistant quartzites and marbles. Underlying all of these more or less horizontal rock formations are even older rocks which have been subjected to multiple phases of metamorphism—alteration by heat and pressure—to form the real foundation or basement of the Sahara. It is fitting to start with these highly altered ancient rocks because they provide the geologic foundation of the Sahara and their influence resonates to this day.

Our planet is about 4.6 billion (4600 million) years old. Much of the Sahara is underlain by rocks that date back to between 2500 and 500 million years ago (see table 1.1 for a geological timescale). These rocks have been greatly altered by extreme heat and pressure and have long been known by the very general term Basement Complex.[3] The Basement Complex rocks have undergone periodic intervals of uplift and deformation followed by intervals of prolonged erosion. They are overlain by more or less undeformed sedimentary rocks that were laid down by water, wind, and ice. This sedimentary cover is up to ten kilometres

TABLE 1.1. The geological timescale (ages are in millions of years)

Cenozoic	Pleistocene	2.6–0.01
	Pliocene	5.3–2.6
	Miocene	23–5.3
	Oligocene	34–23
	Eocene	56–34
	Palaeocene	66–56
Mesozoic	Cretaceous	145–66
	Jurassic	201–145
	Triassic	252–201
Palaeozoic	Permian	299–252
	Carboniferous	359–299
	Devonian	416–359
	Silurian	444–416
	Ordovician	488–444
	Cambrian	542–488
Precambrian	Proterozoic	2500–542
	Archaean	4000–2500
	Hadean	4540–4000

Source: Simplified and adapted from Geological Time Scale, Geological Society of America, 2009, and F.M. Williams (2016). *Understanding Ethiopia: Geology and Scenery.* Springer, Dordrecht, Figure 2.1.

thick and occupies well over half of the present Sahara. As a result, Basement Complex rocks are only visible at the surface of the Sahara in about 15 percent of its total area. The contact between Basement Complex and sedimentary cover is often very sharp and is sometimes evident as a line of springs, some still active, some long dry.

Dating the Basement Complex rocks has been difficult. This is because during each successive phase of mountain building and rock deformation, heat and pressure act on the rocks to reset their apparent age to that particular interval. Ages obtained for these ancient mountain-building or orogenic events range from greater than 2600 million years to about 650–550 million years, with a younger orogenic event bracketed between roughly 350 and 200 million years ago.[4]

In addition to the vertical earth movements that culminated in uplift in different regions of the Sahara, there were also slow but important horizontal movements of the continental crust. Between

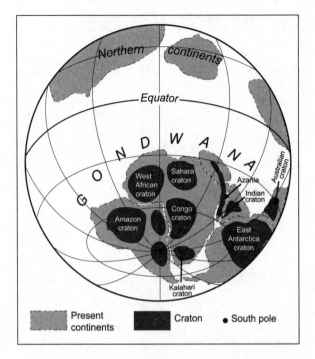

FIGURE 1.1. The super-continent of Gondwana at about 420 million
years ago. Adapted from Springer Nature: F.M. Williams (2016).
Understanding Ethiopia: Geology and Scenery. Springer,
Dordrecht. © Springer International Publishing Switzerland.

about 650 million and 500 million years ago, great slivers of former
oceanic crust and chunks of former continents joined together to form
a super-continent known to geologists as Gondwana[5] (fig. 1.1). The as-
sembly of Gondwana was complete by about 420 million years ago. I
discuss the movements of what geologists refer to as the African plate
or African tectonic plate in greater detail in chapter 4, so only a few key
points need to be considered here. As the Gondwana super-continent
began to break apart some time later, there were vigorous earth movements in the western Sahara and the Atlas Mountains (map 1) associated with the opening of the North Atlantic. Later separation of South
America from Africa led to opening of the South Atlantic. By about

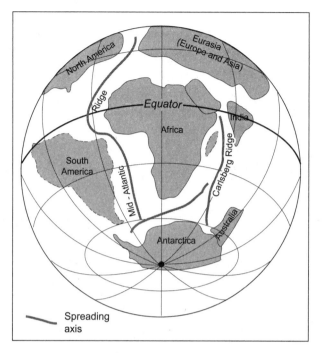

FIGURE 1.2. Gondwana at about 60 million years ago, following the
separation. Adapted from Springer Nature: F.M. Williams (2016).
Understanding Ethiopia: Geology and Scenery. Springer, Dordrecht.
© Springer International Publishing Switzerland.

240 million years ago, the splitting up of the original Gondwana super-
continent was under way (fig. 1.2). It is entirely possible that the separation
took place along zones of ancient crustal weakness, which we describe
in the next section.

Ancient Structures Resonate Today

Although mostly invisible today, the Basement Complex rocks have
long exerted an influence out of all proportion to their present visible
outcrops. This is because certain zones of crustal weakness that developed within the Basement Complex rocks during past intervals of

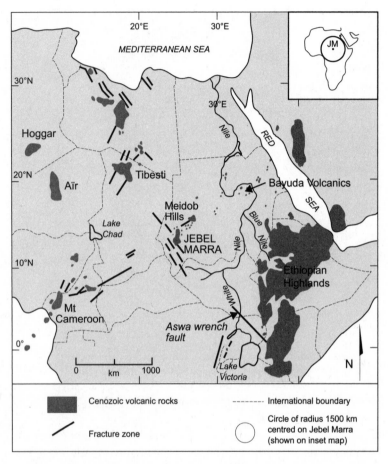

FIGURE 1.3. The location of Jebel Marra volcano in relation to major tectonic lineaments in the Sahara. After M. Williams (2019). *The Nile Basin: Quaternary Geology, Geomorphology and Prehistoric Environments*. Cambridge University Press, Cambridge and New York, fig. 13.1. © Martin Williams. Reproduced with permission of the Licensor through PLSclear.

deformation and uplift have been reactivated at intervals ever since.[6] For example, a belt of volcanoes and ring-complexes runs from the volcanoes in the Gulf of Guinea northeast from the Cameroon volcanic mountains and across northern Sudan (fig. 1.3). These zones of crustal weakness are sometimes simply called lineaments. Another zone of crustal weakness runs from Somalia through South Sudan (map 2) northwest through Tibesti volcano to the Mediterranean coast in Libya. Jebel Marra volcano in western Sudan lies at the intersection of these two lineaments[7] (fig. 1.3). Another major lineament runs north through the Hoggar Mountains. The persisting influence of these ancient lineaments is clearly shown in the great angular bends in the course of the present-day river Nile in northern Sudan, as well as in the orientation of the Red Sea, the Gulf of Aden, and the Ethiopian Rift (fig. 1.3).

Why Are Parts of the Sahara So Flat?

Anyone who has travelled across the Sahara, whether from the Libyan coast southwards to the Sudan border, or from Algeria southwards into Niger or Mali (map 2), is immediately struck by the remarkable flatness of much of the landscape. Examples include the Tademaït sandstone plateau or *hamada* in central Algeria (map 1), with its cover of dark gravel, or the vast Calanscio *Serir* of northern Libya (map 1), with its curious pebble cover that merges southwards into a sand sea of the same name—the Calanscio Sand Sea (see chapter 5, fig. 5.1). The Arabic term *hamada* refers to a rocky plateau devoid of sand, and the Arabic term *serir* denotes a level or gently undulating plain with a more or less continuous gravel cover.

Crossing these vast, flat, empty wastes, whether on foot or by camel, requires a high degree of physical and mental toughness. Small wonder that English author Geoffrey Moorhouse[8] called it 'The Fearful Void' and became quite demoralised in his ultimately unsuccessful attempt to walk across the Sahara. The men who traded the blocks of salt[9] mined from the salt pans at places like Borkou (map 1) in the southern Sahara across the desert were tough, stoical, and fatalistic ('Men of Salt'), but at least they had the prospect of some reward for their labour. For the wretched slaves who were harried along the infamous *Darb al Arba'in* (or Forty Day route), the journey would have seemed endless. To this day the skeletons of camels and humans lying half hidden in the sand attest to the severity of this long walk from northwest Sudan to Dakhla Oasis (map 3) in Egypt (map 2) and on to the Nile.

Every desert land has its own vernacular terms for these landforms. In Australia, *serir* are known as *gibber* plains from the Aboriginal word *gibber* for stone or gravel, and *hamada* are known as stony tablelands. Why, then, is so much of the Saharan landscape so flat? There are a number of reasons. At intervals in the past, the Sahara was invaded by the sea and marine sediments were laid down across the landscape, particularly during Cretaceous times between 145 million and 66 million years ago. These sediments filled preexisting topographic depressions in the land and formed a mantle of nearly horizontal limestones, sandstones, and mudstones. Rivers later cut into these marine sedimentary rocks and eventually formed wide valleys separated by steep cliffs bounding what were now quite extensive plateaux. The surfaces of these plateaux were not subjected to much erosion and so remained fairly level unless later earth movements occurred.

Rivers flowing down from high mountains like the Hoggar and Tibesti (map 1) deposited substantial quantities of sand and gravel in the low-lying depressions surrounding the mountains during about the last 30 million years. Over several million years any irregular bedrock surfaces would become buried beneath a thick layer of alluvial sediments sloping gently down from the uplands. Lakes large and small occupied hollows in the landscape during wetter climatic intervals.[10] When the lakes dried out, the former lakebeds formed flat surfaces of fine-grained sediments. Some of these now dry lake floors extended over a large area, as in the Chad Basin[11] (map 1) and in the Fezzan region of southwest Libya[12] (map 1). But the main reason much of the Sahara is so flat today is because many parts of the Sahara have remained very stable for a very long time, a topic we come to in the next section.

Stability and Erosion

Many parts of the Sahara have been more or less stable for more than 500 million years, with earth movements and volcanic activity being quite localised. I discuss this in chapter 2. Because much of the Sahara has been stable for so long, prolonged erosion has reduced much of the landscape to a very gently undulating erosion plain. During times when the sea invaded the Sahara, these gently sloping erosional plains were blanketed beneath a thin layer of marine sediments, as in northern Egypt. Rivers later flowed across the marine plains as the sea withdrew, depositing gravel, sand, and silt. These sediments were later buried in their turn and compressed to form the sandstones, mudstones, and conglomerates of the Mesozoic Nubian Sandstone Formation[13] that today covers much of the eastern Sahara. Some areas in central Sudan previously thought to be Nubian Sandstone of Cretaceous age may in fact be much younger[14] and on the basis of fossils found within them, appear to be of Cenozoic age, that is, less than 66 million years old. Subsequent uplift and erosion converted these sedimentary rock formations to plateaux, as we saw in the preceding section. The surface of these *hamada* or stony plateaux is still very flat today. In chapter 2, I provide more detail about the depositional history of the Sahara and the most recent phase of uplift and volcanic activity that gave rise to the Saharan uplands we see today.

CHAPTER 2

Birth of the Sahara

There is no novelty in the idea that a mountainous region of interior drainage may be reduced to a plain by the double process of wearing down the ranges and filling up the basins.

W.M. DAVIS, *THE GEOGRAPHICAL CYCLE IN AN ARID CLIMATE*

Legacy of the Past

In chapter 1 we saw that although much of the Sahara is covered in usually quite gently sloping sedimentary rocks, which range in age from only a few million years to more than 500 million years, it is the continuing influence of the ancient Basement Complex rocks beneath these sedimentary rocks that remains important to this day. The Basement Complex rocks were repeatedly folded, faulted, and metamorphosed during the long period of mountain building between about 850 million and 550 million years ago known to geologists as the East African Orogeny.[1] During this time there were several phases of widespread erosion as well as several phases of intrusion of coarse-grained igneous rocks like granite. The coarse crystals in granite denote slow cooling at depths between about five and ten kilometres below the surface. *Intrusion* means that the originally molten rocks penetrated upwards into preexisting, more solid rocks but never reached the surface. Volcanic rocks, in contrast, are *extruded* onto the surface as molten rocks, like basalt lavas, which then cool rapidly and have very small crystals indicative of rapid cooling. During later stages of intense heat and pressure from overlying rocks, the granites were often metamorphosed to gneisses.

There are several reasons why the rocks formed during the time of the East African Orogeny are important in our story of the gradual evolution of the Saharan landscapes. Under the wetter-than-present conditions that prevailed during much of the past 500 million years, these rocks became weathered to form fairly fertile soils that could support a dense cover of tropical forest and woodland. These rocks also contain an abundance of precious and semi-precious minerals such as gold and diamonds that have been attractive to humans from prehistoric times onwards. These ancient rocks have also exerted a more direct influence on the orientation of big rivers like the Nile, the location of large volcanoes like Jebel Marra in northwest Sudan, and the alignment of volcanoes, ring-complexes, and granite intrusions which often follow zones of weakness or lineaments in the Basement Complex rocks that were re-activated at intervals during the last 500 million years. These same lineaments controlled the location of rift valleys and closed sedimentary depressions that now lie buried beneath many thousands of metres of younger sediments, such as the oil-bearing rocks of South Sudan, Libya, and Algeria. In the Sahara, the influence of the past is never far, and its all-pervading legacy is clearly visible today.

In the following sections I will show how what is now the Sahara has been subjected to prolonged periods of uplift, erosion, submergence beneath ice caps, flooding from the sea, and widespread volcanic activity. The climate prevalent across the Sahara has varied from hot, wet equatorial to cold and dry. During the long and mostly wetter interval between 250 million and 66 million years ago, dinosaurs roamed widely across the Sahara. Fossil remains of some of the forests through which they roamed are still in evidence today in the shape of silicified tree trunks that litter many parts of the desert from east to west. The repercussions from a huge asteroid impact 66 million years ago brought an end to the age of dinosaurs and saw the start of the age of mammals. In more recent times, prehistoric hunter-gatherers have made opportunistic use of the fragments of fossil wood in fashioning stone tools when more suitable rocks were not locally available.

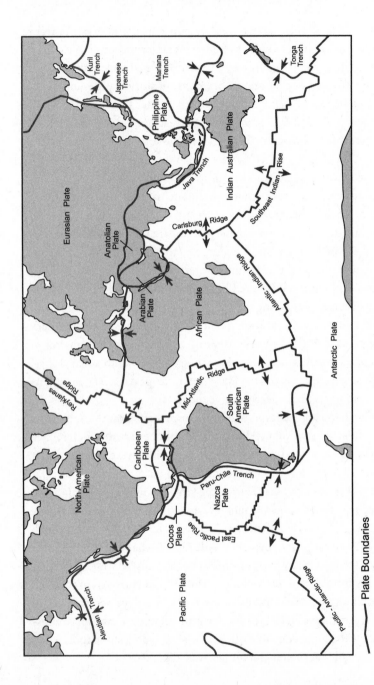

— Plate Boundaries

FIGURE 2.1. Map showing the major present-day tectonic plates. After M. Williams et al. (1998). *Quaternary Environments*, 2nd edition. Arnold, London, fig. 2.1. © Informa UK Limited. Reproduced by permission of Taylor & Francis Group through PLSclear.

When Continents Collide

The outer portion of planet earth is called the crust. Beneath it lies the mantle, which is about 3000 kilometres thick and overlies the molten core of the earth. The crust beneath the oceans consists of dense volcanic rocks and is between five and eight kilometres thick. The lighter crust beneath the continents consists mostly of a whole variety of different types of rock (igneous, metamorphic, sedimentary) and is generally up to thirty kilometres thick but can be twice that thickness beneath very high mountains. The crust consists of what are called *tectonic plates*, which are gently curved slabs of solid rock made up of both continental and oceanic crust.[2] At the present time there are seven major tectonic plates: the African, North American, South American, Antarctic, Pacific, and Indo-Australian plates (fig. 2.1) as well as a number of much smaller plates. These plates are in constant slow motion, sometimes moving apart and sometimes colliding with each other.

In the final stages of the East African Orogeny, between about 650 and 500 million years ago, what geologists call the super-continent of Gondwana was finally assembled from a diverse array of tectonic plates, including those of Africa, Antarctica, South America, and India-Australia (see chapter 1, fig. 1.1). When two tectonic plates collide, the plate that is the denser of the two sinks beneath the lighter plate, causing uplift and a surge of mountain building (fig. 2.2). A good example is the collision between what is now India and the main continental mass of Asia which began about 45 million years ago.[3] The plate on which India now lies was denser than the Asian plate and so slid slowly beneath the southern edge of the Asian plate, causing the rocks above to become folded and compressed, leading to uplift of the Tibetan Plateau and formation of the Himalayas.

The convergence of the four main tectonic plates that formed Gondwana was likewise associated with compression, folding, and faulting of the rocks and the gradual formation of very high mountains. Erosion today is highest in mountainous regions with high rates of precipitation. So, for instance, before they were dammed and regulated, four big rivers that drain the Himalayas today—the Ganges, Brahmaputra, Indus, and

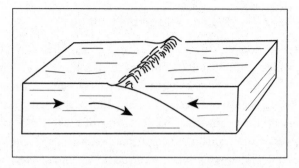

FIGURE 2.2. Collision of two tectonic plates leading to uplift
and mountain building.

Yangste—together provided nearly one-fifth of all the sediments flowing into the oceans. The rivers that eroded the mountains of Gondwana
would likewise have deposited large volumes of sediment into the
oceans surrounding the super-continent. Some of the sediments would
also have been spread across the Saharan lowlands to form the Cambrian sandstones that crop out in a few places across the modern Sahara.
The Cambrian (see the geological timescale in chapter 1, table 1.1) was
also a time when multi-celled plants and animals became abundant
across the world, ushering in an entirely new era in which the composition of the atmosphere was now being increasingly controlled by life on
earth as a result of plants absorbing carbon dioxide from the atmosphere during photosynthesis and emitting oxygen during daylight
hours. Photosynthesis is the process whereby green plants use energy
from sunlight to manufacture glucose from carbon dioxide and water
and emit oxygen as a by-product.

A Very Strange Sort of Mountain

Scattered across certain parts of the Sahara are strange and unique
mountains known as 'ring-complexes'. These ring-complexes[4] have a
central core of granitic rocks flanked by circular bands of volcanic rocks
(fig. 2.3). Seen from the air, each one looks like a gigantic bull's-eye
made up of steep, narrow concentric ridges separated by narrow valleys.

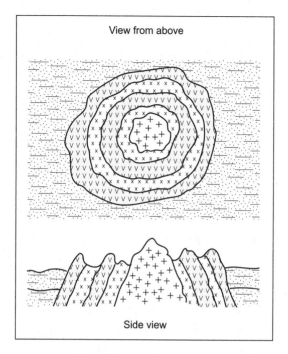

FIGURE 2.3. Schematic cross-section through a ring-complex
and view from above.

Looking down from above, the overall shape is more or less circular. The
Saharan ring-complexes are of considerable interest to archaeologists
because they were attractive localities for prehistoric hunter-gatherers
and prehistoric pastoralists. Water was the main reason for this attraction. Springs issuing from the flanks of some of the larger ring-complexes
like Jebel 'Uweinat in southeast Libya (map 1) provided permanent
fresh water for drinking. Small lakes or wetlands were often trapped
along the valley floors between the ridges and were host to several species of fish, including Nile perch, as well as crocodiles, turtles, and hippos, as at Adrar Bous ring-complex in the Ténéré Desert east of the
northern Aïr massif (map 1). Some of these lakes were fed by runoff
from the mountain, some by groundwater during wetter climatic intervals, and some by a combination of runoff and groundwater.

Years of careful mapping in the 1960s by French geologist Russell Black and his colleagues have shown that more than three-quarters of the Aïr Mountains consist of what geologists have called Younger Granite ring structures intruded through the Precambrian basement rocks.[5] These are all ring-complexes. The Aïr has more than twenty-five such structures, which are between two and thirty kilometres in diameter, with an average diameter of about fifteen kilometres. Because they were sometimes intruded into older sedimentary rocks, some ring-complexes have a cover of resistant sedimentary rocks, but in many cases this former cover has been entirely removed by erosion so that the highest peaks in the ring-complex often consist of granite. Extending from northern Nigeria to northern Niger there are sixty individual ring-complexes aligned in a narrow belt from south to north. In Nigeria, the ring-complexes range in age from early to mid-Jurassic (see chapter 1, table 1.1) or from about 200 million to 175 million years ago, at a time when dinosaurs roamed the Sahara. The Aïr ring-complexes are older and were formed during Silurian and Devonian times (440–360 million years ago: chapter 1, table 1.1). Adrar Bous is the most northerly of the Aïr ring-complexes and has an early Silurian age of 430 million years.[6] The ages of the ring-complexes become progressively younger as you go south. One possible explanation for this age trend is that during its slow northward movement, the African tectonic plate passed directly over one or more stationary hot spots[7] that were present in the earth's mantle during this long interval of geologic time (fig. 2.4).

There is another group of granite intrusions and associated ring-complexes that runs diagonally from southwest to northeast from the Cameroon Mountains in West Africa across the southern Sahara through southern Libya, northern Sudan, and northern Eritrea to the Red Sea Hills (see chapter 1, fig. 1.3). Here again, erosion of the softer and more weathered rocks has created narrow valleys flanked by concentric ridges of resistant rocks. Alluvial fans criss-crossed by ephemeral stream channels line the margins of the valleys and during wetter times, flowed into now dry lakes.

During favourable times in the last one million years, from Early Stone Age times onwards, people have visited and lived in these valleys.

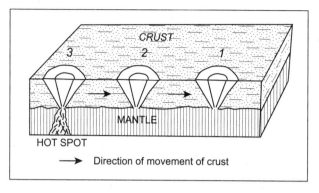

FIGURE 2.4. Ring-complex formed as a result of the African tectonic
plate moving over a hot spot in the underlying mantle.

The later arrivals made imaginative use of smooth rock surfaces in certain rock shelters to paint lively scenes of daily life, including prehistoric
hunting scenes with hunters and their dogs and scenes showing great
herds of brindled Neolithic cattle. Jebel 'Uweinat ring-complex alone
has well over a thousand recorded rock paintings and rock engravings.[8]
Some of these paintings are remarkably lifelike. One shows a goat on its
hind legs browsing on a tree that looks very similar to *Maerua crassifolia*,[9] a tree that in northern Sudan, camels (and no doubt goats) have
always found very attractive as food. The solitary tree at the outlet of the
main wadi at Jebel Arkenu, a ring-complex forty kilometres northwest
of Jebel 'Uweinat, is also a *Maerua crassifolia* tree.[10]

Under the Ice

Mountain building was accompanied and followed by prolonged erosion and sediment deposition. It was during these times that sandstones
like the Mesozoic Nubian Sandstone Formation were laid down across
much of the Sahara and the then adjoining region of Arabia. Today the
Nubian Sandstone is the largest freshwater aquifer in the world and is
widely used across the Sahara, especially in deep wells dug into the
sandstone. The two most recent phases of recharge of the groundwater

within the Nubian Sandstone Formation were centred on about 125,000 and 10,000 years ago;[11] that included a long interval with no recharge synchronous with the cold, dry phase 20,000 years ago known as the Last Glacial Maximum because ice caps then covered much of North America and western Europe.

Interspersed between these long periods of erosion and sediment deposition there were occasional intervals of extreme cold during which ice caps covered much of what is now the Sahara. One such glaciation was the Ordovician glaciation[12] of 460–440 million years ago (chapter 1, table 1.1) during which glacial deposits accumulated to substantial thicknesses between Mauritania and Arabia. The ice scoured down to bedrock. The resulting glacially scoured rock pavements in the Hoggar region of the central Sahara have only recently been exposed as a result of erosion of the protective cover of overlying sedimentary rocks. They appear so fresh[13] that they can easily be mistaken for the most recent glacial erosion of only 20,000 years ago. Once the ice melted, sea levels rose worldwide and flooded the previously exposed edges of the continents.

There was another major glaciation between about 360 and 260 million years ago during which much of the super-continent of Gondwana was once again beneath the ice. This glaciation is known as the Permo-Carboniferous glaciation (chapter 1, table 1.1). The end of the Permian about 250 million years ago was accompanied by widespread extinction[14] of almost all marine life and a very high proportion of terrestrial life. This mass extinction event had nothing to do with the preceding glaciation but may have been caused by a phase of extreme volcanic activity and the release of substantial volumes of acid rain into the oceans and onto the land.

The last and most recent phase of glaciation began with the accumulation of permanent ice caps in Antarctica about 34 million years ago, followed much later by the formation of ice caps over North America and northwest Europe starting about 2.6 million years ago.[15] This was the time of the Quaternary ice ages and the time when prehistoric humans first appeared. The earliest glacial-interglacial cycles were relatively short and not very intense, each lasting only about 20,000 years.

From about 0.7 million years onwards these cycles were longer and more intense, with each cycle lasting about 100,000 years but interspersed with shorter fluctuations. During the peak of the last glacial cycle, known to glacial geologists as the 'Last Glacial Maximum', which lasted from about 26,000 to 19,000 years ago, the Atlas Mountains (map 1) were covered in permanent snow and supported numerous valley glaciers.[16] The upper treeline was about a thousand metres lower than today, temperatures were 8–12 degrees Celsius cooler. In the higher parts of the Hoggar Mountains, many rocks were broken apart by water freezing in cracks in the rocks and wedging them apart. The overall climate at this time across the Sahara was cold, windy, and very dry, which would have made life hard for tropical plants and animals not adapted to such extreme conditions. Many previously stable dunes were now active, so that the area of desert was greatly enlarged along its southern margins.[17] It is probable that many of the Late Stone Age hunter-gatherers abandoned the Sahara at this time.

Roaming Dinosaurs and Petrified Forests

The word 'dinosaur' comes from two Greek words: *deinos* meaning terrible and *sauros* meaning lizard, and it was coined by Sir Richard Owen in 1841.[18] It takes a considerable effort of the imagination today to visualise dinosaurs wandering across the Sahara. The contrast between the present waterless and barren desert and the former presence of dinosaurs large and small, some herbivorous, some carnivorous, seems too much to accept, but the fossils are there and tell us a tale of a time when the Sahara was able to support rivers and lakes, swamps and wetlands, and a luxuriant plant cover. Some dinosaurs were as big and heavy as a small truck, which helps to explain Owen's invented term.

It was during the Mesozoic Era, which lasted from 250 to 66 million years ago, that dinosaurs roamed the earth. The Mesozoic could aptly be called the 'dinosaur era'. Although every continent has yielded dinosaur fossils, during certain times in the Mesozoic, dinosaurs were especially common in North Africa. The Mesozoic starts with the Triassic Period (250–200 million years ago), followed by the Jurassic Period

(200–145 million years ago), and ends with the Cretaceous Period (145–66 million years ago). The Triassic was generally hot and dry worldwide, with large deserts in North America and Europe, so that the Triassic dinosaurs were adapted to living in harsh conditions. The climate was generally wetter during the Jurassic, which is sometimes called 'the age of reptiles' because they were so abundant. The oldest Stegosaur in the world—*Adratiklit boulahfa*—comes from Middle Jurassic rocks in Morocco[19] (map 2).

Gondwana was beginning to split apart in the late Jurassic so that more and more of the land was flanked by ocean water, which brought moisture to the adjoining land. South America separated from Gondwana during this time and went on to develop its own unique fauna.

The climate was wet again during the Cretaceous, especially during the Middle Cretaceous, when the Sahara had a distinctly equatorial climate—hot, humid, with minimal daily temperature fluctuations.[20] Dinosaurs were especially numerous in North Africa during the Upper Cretaceous (100–66 million years ago), and well-preserved dinosaur teeth have been recovered from Egypt, Algeria, Tunisia, and Niger (map 2). The fish-eating *Spinosaurus aegyptiacus* lived near rivers and, like crocodiles, had teeth well adapted to catching fish.[21] A newly discovered Upper Cretaceous sauropod from Dakhla Oasis in the Western Desert of Egypt (map 3), estimated to be between about 94 and 66 million years in age, appears to show some evidence of sauropod dispersal between Europe and Africa.[22] If correct, this would mean that the Saharan dinosaurs of that time had not evolved in isolation.

Dinosaurs roamed across what is now the Sahara for almost 200 million years. The dinosaur era ended, quite literally, with a bang. A huge asteroid collided with the earth about 66 million years ago and created an enormous crater—the Chicxulub crater—on the Yucatán Peninsula in Mexico.[23] The secondary effects of this enormous impact included forest fires and acidic oceans, and it caused another mass extinction event on earth, of which the demise of the dinosaurs was a significant part.

By a curious quirk of geology, relics of the vast forests and woodlands through which Saharan dinosaurs wandered a hundred million years ago are preserved to this day in the form of silicified tree trunks. Many parts of the Sahara are littered with the remains of these former forests, and some of the fossil tree trunks are fifty feet (15 metres) long and twenty inches (50 centimetres) in diameter. Some of the trees are still in growth position and must have been buried quite rapidly in sand before all the cells of the trees were replaced with silica. The trees have been preserved because weathering was very active during the Cretaceous, when the climate was hot and wet. Dissolved silica released during weathering of any rocks rich in silica, including the quartz sands and gravels that later became the Nubian Sandstone, was carried downstream by rivers and later precipitated in the near surface sediments of the former floodplains. There were repeated phases of silica precipitation so that trees growing on the sandy floodplains were silicified and buried beneath younger layers of sand.

The Nubian Sandstone Formation in the deserts of northern Sudan and southern Libya sometimes has a very resistant capping made up of iron and silica, which effectively preserves the underlying softer sandstones and shales from erosion. In due course, these resistant sandstone plateaux become undercut and the slopes retreat, releasing their treasure trove of fossil wood. Fossil wood is especially common today in areas where the Nubian Sandstone has been almost completely eroded away, leaving little doubt that the fossil wood came from within the sandstones. In a few places the tree trunks have only recently been exhumed, and remain upright. These have been called petrified forests, like the fossil trees in the famous Petrified Forest National Park in Arizona, or those at Lesbos in Greece, or the smaller patches of fossil forest near Khartoum in Sudan and Cairo in Egypt. The French soldier-naturalist Charles Sonnini de Manoncourt (1751–1812) travelled widely in Egypt in 1777–78 and brought back fragments of petrified wood from the soda lake in Wadi el Natrun in the desert of northwest Egypt.[24]

Our prehistoric ancestors made good use of the fossil wood scattered across the surface of the Sahara. With some careful trimming, it produced serviceable stone tools. Today we can find such tools scattered across a wide swathe of country extending from southern Algeria across

Niger and southern Libya to the Western Desert of Egypt and the Nubian Desert of northern Sudan (map 2).

Beneath the Sea

The Cretaceous (145–66 million years ago) was a time of very high sea levels across the planet, with occasional widespread flooding of the continental lowlands. One reason for the high sea levels at these times was a resurgence of plate movements and of increased volcanic activity along the mid-ocean ridges where plate movement was initiated.[25] As more lavas poured over the ocean floors, sea level rose. Some of the marine flooding was aided by faulting and the formation of coastal rift valleys that reached below sea level, as in West Africa. For a brief period sometime between about 90 and 100 million years ago there was a continuous sea link between the South Atlantic and the Tethys Sea in the north.[26] The Tethys was a vast ocean extending across to what is now northern India. The maximum extent of flooding was about 94 million years ago. At this time only the uplands and the higher parts of the Precambrian Saharan shield rose above the sea.

Marine Cretaceous sediments can be seen today at elevations approaching 2000 metres in the northern Aïr Mountains,[27] indicating extensive faulting and significant uplift once the seas had withdrawn. A deeply buried and downfaulted rift valley east of the Aïr contains marine Cretaceous rocks.[28] At the peak of these Cretaceous marine invasions, much of the Sahara lay beneath the sea, as did the Ogaden region of Ethiopia and much of Somalia (map 2). A further marine invasion between about 90 and 66 million years ago reached as far as the Western Desert of Egypt, but after this there were no further invasions of the sea of any consequence. The limestones that bear witness to these former marine incursions are important aquifers today in places like Dakhla Oasis in the Western Desert of Egypt (map 3), which has been occupied throughout human prehistory and was a minor desert outpost of the Roman Empire, as shown by its still well-preserved buildings and their engravings.

Birth of the Saharan Uplands

As the African plate moved northwards and began to encroach on the Eurasian plate, parts of the earth's crust buckled and were thrust upwards between about 370 and 290 million years ago. It was during this long interval of crustal deformation and uplift that the Atlas Mountains in northwest Africa and the Alps in Europe first came into existence, although there were many phases of uplift and erosion following that time. Some parts of the previously very stable Precambrian rocks were also affected, resulting in localised uplift and volcanic activity in places like the Hoggar, Tibesti, and Jebel Marra (fig. 2.5). Further uplift and volcanism were to occur later, in the Cenozoic Era, which spans the last 66 million years and is popularly known as the age of mammals.

The global climate fluctuated quite wildly during the course of the Cenozoic,[29] but the overall trend was one of progressively cooler conditions across our planet. During the first 30 million years of the Cenozoic (Palaeocene: 66–56 million years ago; Eocene: 56–34 million years ago), the far northwest of the Sahara began to show the first signs of aridity[30] while the south-central Sahara was still experiencing a hot, wet equatorial climate and was at least intermittently covered in tropical rain forest. This was a time of extreme weathering in the regions under forest, resulting in the formation of a surface layer of deeply weathered rocks across much of the southern and central Sahara. This period of tectonic stability and equable climate was followed by a period of uplift and erosion that lasted for nearly ten million years and gave rise to a major erosion surface[31] extending from the Negev Desert of Israel through the Sinai Desert and across North Africa to the Hoggar.

Continuing uplift was accompanied by climatic desiccation, widespread erosion, and gradual exposure of the irregular weathering front (fig. 2.6). The major elements of the Saharan landscape came into being at about this time, and this may well denote the birth of the Sahara as a desert. This is not to say that the Sahara was as arid as it is today. It was not. Huge rivers flowed across the Sahara at intervals during the Miocene between about 20 million and 5 million years ago. Big

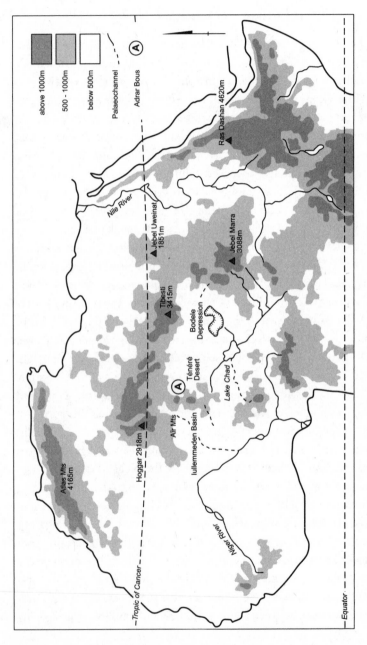

FIGURE 2.5. Map of the major Saharan uplands. After M. Williams (2014). *Climate Change in Deserts: Past, Present and Future*. Cambridge University Press, Cambridge and New York, fig. 18.3. © Cambridge University Press. Reproduced with permission of the Licensor through PLSclear.

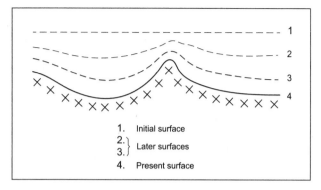

1. Initial surface
2. } Later surfaces
3. }
4. Present surface

FIGURE 2.6. Schematic diagram showing exposure of the irregular
weathering front in deeply weathered Saharan rocks as a result of erosion.

rivers flowing northwards from the northern edge of the Chad Basin cut
wide valleys between Tibesti to the west and three large Nubian Sandstone
plateaux to the east. The highly dissected margins of these sandstone plateaux suggest that small streams may have flowed across them and may
have acted as tributaries to the larger trans-Saharan rivers of that time.
Fossil remains of plants and animals show that the Chad Basin was covered
in a mosaic of wetlands, savanna grasslands, and woodlands during the
late Miocene,[32] consistent with a somewhat wetter climate in the headwaters of the rivers flowing north from that region across the Sahara.

Miocene and later volcanic activity in the Hoggar, Tibesti, and Jebel
Marra accompanied and most likely also preceded uplift of the underlying
basement rocks, amounting to about a thousand metres of basement uplift in the Hoggar and Tibesti and about five hundred metres in the case
of Jebel Marra.[33] About 8000 cubic kilometres of rock were erupted during the formation of Jebel Marra and about 3000 cubic kilometres at
Tibesti. Tibesti (3415 m), Jebel Marra (3088 m), and the Hoggar (2918 m)
are the three highest peaks in the Sahara and have benefitted from more
frequent and more abundant rainfall than elsewhere. As a result, they have
long acted as refuge areas for plants and animals left stranded during drier
climatic intervals. On the southern footslope of Jebel Marra (figures 2.7
and 2.8) there are very well-preserved fossil leaves of the oil palm *Elaeis*

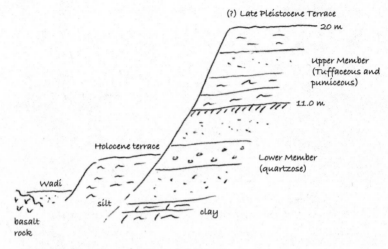

Jebel Marra southern footslope (19/1/76)
(Umm Mari section D)

FIGURE 2.7. Sketch of alluvial terraces bordering a valley cut into the southern slopes of Jebel Marra volcano, Darfur Province, western Sudan (19 January 1976). The terraces are evidence of alternating vertical erosion and fluvial deposition.

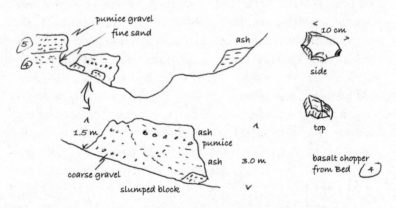

Umm Mari, section G, detail (20/1/76)
southern foot slopes, Jebel Marra, Darfur, Sudan

FIGURE 2.8. Sketch showing detail of slumped block on the side of a valley cut into the southern slopes of Jebel Marra volcano, Darfur Province, western Sudan (20 January 1976). Embedded in Bed 4 were Early Stone Age artefacts with a probable age range of about 0.8–1.5 million years, as well as oil palm (*Elaeis guineensis*) and *Combretum* leaf fossils, indicative of a very much wetter climate at that time than the present semi-arid climate.

guineensis which are found in reworked volcanic ash beds that contain Early Stone Age stone tools with a probable age of about a million years.[34] The oil palm grows today in rain forest extending for up to three hundred kilometres inland from the coast of West Africa. The climate must have been very much wetter, with permanent rivers flowing south from the volcano to enable the oil palms to migrate from West Africa along the rain forests flanking the rivers to finally reach Jebel Marra.

As these events were taking place in the central and southern Sahara, even more dramatic events were already under way to the east, with uplift in Ethiopia and opening of the Red Sea. Sporadic volcanic activity has continued until relatively recently in the deserts of northern Sudan and southeast Libya, where volcanic plugs consisting of very fresh columnar basalt have pushed through the Nubian Sandstone to form isolated hills a hundred metres or more in height. The presence of occasional Middle and Late Stone Age artefacts on the summits of these volcanic hills suggests that they were probably used as lookouts by prehistoric hunters.

In chapter 3 we move much closer to the present and look at a time when the Sahara was once again a land of rivers and lakes, attracting prehistoric hunter-gatherers to settle close to the lakes and hunt the hippos, crocodiles, and turtles that lived in them.

CHAPTER 3

Hippo Hunters of the Sahara

The Garamantes hunt ... in four-horse chariots

HERODOTUS, *THE HISTORIES*

When the Sahara Was Green

Throughout its long history the Saharan climate has oscillated between wet and dry.[1] The last time that the Sahara was significantly wetter was between about 15,000 and 5000 years ago.[2] During that long interval of mostly much more humid conditions, a whole series of inter-connected rivers and lakes allowed free passage across the Sahara to plants and animals.[3] The vegetation for much of that time was very reminiscent of the savanna woodlands and savanna grasslands[4] of East Africa today with their herds of antelopes, elephants, and giraffes, together with occasional rhinos and the ever-present bands of predatory wild dogs, jackals, hyenas, and lions. A savanna ecosystem is one where the tree canopy remains open so that sunlight can penetrate between the trees to allow grasses and smaller shrubs to flourish. It is characteristic of the seasonally wet tropics. Where grasses are more abundant than trees it is called a savanna grassland; where trees are more common it is called a savanna woodland.

The fossil remains of these savanna animals have been found across the Sahara and they are also depicted in prehistoric rock engravings and paintings.[5] More impressive still are the fossil remains of aquatic animals large and small. The larger animals include crocodiles, hippos, turtles, and Nile perch (plate 3.1), which are found as fossils in lakes[6] (plate 3.2) across the central and southern Sahara at a time when the Sahara could aptly be described as green. The smaller aquatic creatures include amphibians that were able to traverse the entire desert, and aquatic snails that were confined to permanent rivers and lakes.[7] The fossil shells of these aquatic snails are common in the sediments of former lakes that were scattered across the entire Sahara.

Even more intriguing than the remains of crocodiles and hippos are the tantalising remains of domestic Neolithic cattle[8] which used to graze across what is now bare and waterless desert and first appeared in the Sahara about 7400 years ago. The Neolithic period began about 12,000 years ago in the Near East and marks the end of the Late Stone Age in that region. It refers to the time when humans first domesticated plants and animals and began to use polished stone implements but had not yet discovered how to use metals. Cattle were obviously highly prized by their nomadic pastoral owners because they also appear as multi-coloured paintings on suitable rock surfaces on sites as far removed as the Tassili n'Ajjer[9] (map 1), a sandstone plateau in Saharan southeast Algeria, and Jebel 'Uweinat[10] (map 1), a 1934-metres high mountain sitting astride the present-day frontiers of Libya, Egypt, and Sudan in the far southeast of the now hyper-arid Libyan Desert. Table 3.1 is a summary of the type of evidence used to reconstruct prehistoric environments in the Sahara. In the following sections I discuss how such evidence has been used to reconstruct the kinds of environments that were present in the Sahara during this final wet phase when the desert became green for the last time.

A Land of Lakes and Rivers

In the prologue and introduction to this book, I described my journey to Adrar Bous (plate 3.3), an isolated mountain in the geographical heart of the Sahara. One of the things that became obvious during our first few days at Adrar Bous, when we saw the remains of the former sandy beaches and freshwater snail shells (figures 3.1 and 3.2), was that a relatively deep freshwater lake had once occupied the wide valley located

TABLE 3.1. Evidence used to reconstruct environmental change in the Sahara

Type of evidence
Geology
Soils and fossil soils
Lakes and lake sediments
Wind-blown sediments—desert dust, dunes, sandplains
Glacial deposits (e.g., moraines) and features of glacial erosion (e.g., glacial valleys)
Biology
Fossil pollen and spores; plant macrofossils and microfossils
Vertebrate fossils; invertebrate fossils: mollusca, ostracods, diatoms, insects
Modern population distribution
Archaeology
Plant remains; animal remains, including hominins; cemeteries; rock art; hearths, dwellings, workshops; artefacts: bone, stone, wood, shell, leather

near the southern margin of the mountain and bounded to the south by a long, narrow ridge of rock that was currently becoming concealed beneath large climbing dunes. On a still and quiet night, the eerie reverberations from the avalanching slip face of the dunes sounded like some ghostly distant drummer. In fact, the early French Saharan explorers used to refer to this strange desert music as 'le tambour des dunes' or the drumbeat of the dunes. The local nomads had more picturesque descriptions involving djinns or genies.

The lake had been deep enough to accommodate a rich fauna of fish, reptiles, and aquatic mammals, including Nile perch, catfish, turtles, crocodiles, and hippos, as well as aquatic birds.[11] We found the remains of these in the prehistoric middens that we excavated. Scattered across the Sahara at this time, roughly 9000 years ago, there were abundant lakes and streams, so it is no surprise that French archaeologists referred to this period as 'Le Sahara des Lacs' or 'the Sahara of Lakes'.

Some of these lakes were huge, such as the vastly expanded Lake Chad[12] (which first arose during the Miocene) and the lakes in southwest Libya.[13] Others were large but not huge; while others again were relatively small, including the lake at Adrar Bous, and scattered lakes elsewhere in Niger and Chad as well as in Mauritania, Mali, Algeria, northern Sudan, and the Western Desert of Egypt.[14] Many of the lakes were connected to rivers, allowing plants and animals to migrate from

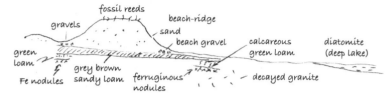

Ferricrete ridge section, Adrar Bous (5/3/70)

FIGURE 3.1. Sketch of sandy beach-ridge and associated lake deposits at Adrar Bous mountain in the Ténéré Desert of Niger, south-central Sahara (5 March 1970). This lake was high between about 11,000 and 8,500 years ago.

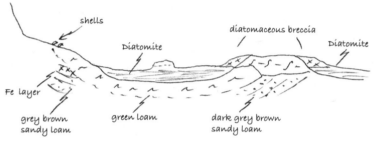

Diatomite II section (schematic)
Adrar Bous (5/3/70)

FIGURE 3.2. Sketch of a simplified cross-section showing lake sediments revealed in soil pits at Adrar Bous mountain in the Ténéré Desert of Niger, south-central Sahara (5 March 1970). The shells mark the highest level of the lake, which was full and fresh between about 11,000 and 8,500 years ago.

the well-watered northern and southern borders of the present Sahara into the now arid interior.

It is hard to imagine that Nile perch, turtles, crocodiles, and hippos could have reached Adrar Bous unless there was a more or less continuous network of river channels flowing from places like the Nile and Chad basins far into the Sahara. Because much of the evidence of these former channels now lies deeply buried beneath desert dunes and sand sheets, it is not easy for the modern traveller to imagine how different things were even as recently (to a geologist!) as a mere 5000 years ago. However, traces of these ancient river channels are clearly visible on satellite imagery[15] and have also been revealed by radar.[16]

The great French Saharan naturalist and polymath Théodore Monod (who has an attractive city square named after him in the Latin Quarter of Paris) noted the presence of relict populations of patas monkeys (*Cercopithecus patas*) and baboons (*Papio anubis*) in sheltered valleys in the Aïr Mountains.[17] He considered that these simian refugees were descended from ancestral populations of monkeys and baboons that had originally migrated from the West African rain forest during wetter times, when the forests were linked to the mountains by streams flanked by gallery forests which extended like long, green fingers deep across the present desert. Once the climate became dry and the streams ceased to flow, the monkeys retreated to their present refuge in the mountains which, being high, receive more rainfall than the surrounding arid plains, and there they remain today: a relict population from wetter times past.

A good example of a present-day desert river is the Awash River which flows from the well-watered Ethiopian Highlands down to the Afar Desert (map 1), where it disappears into a shallow lake at the end of its present channel. All along its present course through the desert it is flanked by often dense thickets of thorn trees and scrub, providing food and shelter to herds of animals that would otherwise be unable to survive the rigours of the surrounding desert. The river itself is home to various species of fish as well as carnivorous crocodiles and herbivorous hippos. Fetching water from the Awash is always exciting!

Similarly, the occasional venerable olive trees[18] found today in sheltered sites in the northern Aïr Mountains and in deep ravines on the slopes of the Jebel Marra volcanic massif (map 1) in northwest Sudan are remnants of a once more luxuriant population of Mediterranean trees that had moved progressively southwards across the once green Sahara, when great rivers flowed south from the Atlas Mountains into the south-central Sahara, traversing much of the present desert, and also enabling humans to move along these drainage channels. Today, the River Nile is the only river that maintains its course through the desert country of the eastern Sahara, losing much of its water during its long course northwards to the Mediterranean Sea.

Recent work by British geographer Nick Drake and his colleagues using satellite imagery and ground-penetrating radar supplemented by local fieldwork has shown how the Sahara was criss-crossed by a series of interconnected rivers and lakes that allowed easy passage across what is now waterless desert for a wide variety of plants and animals during a long humid interval between about 15,000 and 5000 years ago.[19] Different prehistoric groups of hunter-gatherers made use of these waterways in their occupation of vast tracts of the Sahara. In the southern and central Sahara, people equipped with barbed bone points[20] hunted hippos, crocodiles, turtles, and Nile perch in many of the small lakes that were scattered across this vast region. Further north there was less of a focus on harvesting aquatic foods and more on hunting savanna animals using spears armed with stone points as well as bows and arrows. The arrows were tipped with a distinctive type of stone arrowhead that was widespread across the northern Sahara.

It was not continuously wet between 15,000 and 5000 years ago. More arid intervals were interspersed throughout this time. During the drier climatic intervals there were suitable refuge areas available in upland areas as well as in lowland areas endowed with permanent groundwater springs. In some localities, intense deflation immediately upwind of isolated sandstone hills during drier intervals created wind-scoured hollows tens of metres deep. The base of these hollows sometimes intersected the local water table and a small lake came into being. In one such site, located thirty-five kilometres southeast of Adrar Bous, I found embedded in the finely layered silts of the former lake the flattened remains of a fossilised fish, complete with scales, that died more than 5000 years ago.

Prehistoric Feasts and Parasites

Let us return to Adrar Bous and its former lake. The lake deposits preserved there at present and their associated fossil fauna indicate very clearly that there were at least two periods during which the lake was at a high level. The ages for these high lake phases are based on radiocarbon ages[21] on shells, charcoal, and organic remains found in prehistoric middens. Radiocarbon ages are based on measuring the amount of radioactive carbon still present in fossil bones or charcoal. The method can give useful ages for samples up to about 50,000 years old.

The earliest high lake phase at Adrar Bous is dated to between at least 11,000 and roughly 8500 years ago, when the lake shoreline was at about 710 metres above sea level and the lake was about 13 metres deep. Our excavations showed that the lake was a source of food and water for a group of people who practised fishing, hunting, and plant collecting and made some very characteristic small and beautifully fashioned stone tools (or 'microliths') which are still found around the old lake shore. Archaeologist Andy Smith coined the term 'Kiffian' for this prehistoric human culture,[22] a name that has since been widely accepted by many Saharan archaeologists. The word 'Kiffi' means fish in Tamasheq, the language of the Kel Aïr Tuareg living in the mountains today. We named a small rocky hill overlooking the older lake sediments 'Adrar n'Kiffi' or Fish Hill.

The lake also supported an abundant fauna of freshwater snails, including certain species (*Bulinus* and *Biomphalaria*) that are known today to be carriers of the parasitic flatworm *Schistosoma*, agents of a disease popularly known as *Bilharzia* or in medicine as *Schistosomiasis*. The German pathologist Dr Theodor Bilharz (1825–1862) discovered and named this unpleasant parasite when working in Egypt in 1851, and we now know that the disease was no respecter of persons, the abundant eggs of this parasite having been found in the stomach of mummified Egyptian and Nubian pharaohs.[23] The parasite attacks the internal organs of humans and other mammals who drink or paddle in slow-moving or still waters attractive to these aquatic snails. It weakens but does not kill its host. It is entirely likely that the prehistoric lakeside dwellers at Adrar Bous and elsewhere in the Sahara were infested with these parasites, just as many peasant farmers in the tropics who rely on irrigation are likewise afflicted to this day.

Arid Intervals

Sometime after about 8500 years ago, possibly during a cold and very dry spell dated elsewhere in the Sahara at 8200 years ago,[24] the Adrar Bous lake dried out and remained dry for well over a thousand years. We know this because once the lake had dried out, sand blew in from the desert and covered the dry bed of the lake in a layer of wind-blown sand. When the lake eventually refilled, which it did some time before about 6300 years ago, it never returned to its previous level, and at its highest was always about 30 feet (10 m) lower than before and not much more than about three metres (10 ft) deep. The people who came to Adrar Bous and its now diminished lake brought with them a very different lifestyle to that of their predecessors, although many elements in their stone tool kit were not that different. These new arrivals brought domesticated cattle[25] with them and used polished stone axes and adzes. According to the distinguished British archaeologist Anthony J. Arkell (1898–1980), who excavated a number of Mesolithic and Neolithic sites along the Nile in central Sudan more than seventy years ago, the adze was still the favourite tool of traditional Sudanese carpenters and boat-builders.[26] Sites broadly similar to the Adrar Bous Neolithic sites (known to archaeologists by the local term 'Tenerian', derived from the name of the surrounding Ténéré Desert; see map 1) were common across the southern and central Sahara. These Saharan Neolithic cattle herders often and understandably showed a strong preference for mountains with reliable sources of fresh water supplied by shallow groundwater or local springs. One such mountain is Jebel 'Uweinat, located astride the present-day frontiers of Libya, Sudan, and Egypt, and renowned for its abundant rock paintings, many depicting scenes from the daily life of these Neolithic pastoralists.

We do not know how many intervals of aridity afflicted the Neolithic and earlier inhabitants of Adrar Bous during the last 15,000 or so years. The Tenerian Neolithic cattle herders were present at Adrar Bous between roughly 6300 and 4200 years ago,[27] but their occupation may not have been continuous. In northern Sudan, the Swiss archaeologist Matthieu Honegger and his team have spent many years excavating Neolithic occupation sites and a vast Neolithic cemetery not far from the town of Kerma (map 3), south of the 3rd Cataract. They have now accumulated more than a hundred radiocarbon ages on charcoal and shell from these sites and have found an almost total absence of any occupation between 7500 and 7100 years ago and again between 6000 and 5400 years ago.[28] There is a well-known saying in popular law that

'Absence of evidence is not evidence of absence'. True enough, but these two apparent gaps in Neolithic occupation in northern Sudan also happen to be times of much reduced Nile flow[29] as well as times when a number of Saharan lakes dried out.

Not all lakes respond rapidly to regional changes in climate, particularly lakes fed largely from groundwater, which acts as a buffer against short-term changes in rainfall and runoff. However, some of the Saharan lakes appear to have dried out very swiftly and refilled equally swiftly. Such rapid changes in a once reliable water supply would have had a dramatic effect on plants, animals, and human societies dependent upon them: they could either migrate or, perhaps, adapt. If unable to move or adapt, they became extinct. We return to this theme later in this book.

Lakeside Cemeteries in the Desert

In the Ténéré Desert of Niger, about five hundred kilometres south of Adrar Bous and about two hundred kilometres east of the town of Agadès (map 1), there is an important prehistoric site known as Gobero[30] (map 1). This site contains the earliest known cemetery in the Sahara, with more than two hundred human burials spanning about 5000 years between 9700 and 4500 years ago. American palaeontologist Paul Sereno, well known for his many discoveries of Saharan dinosaur fossils, directed the work at Gobero. The site chronology is based on seventy-eight radiocarbon ages obtained on human remains (especially tooth enamel), the fossil fauna, and prehistoric middens. The dunes that contain the burials and surround the former lake are dated using optically stimulated luminescence methods. Luminescence dating tells us the last time that a grain of sand was exposed to sunlight before burial. Since dunes are formed as exposed sand grains are continually being buried by more sand blowing over them, this can tell us the age range over which the dune formed. This technique has yielded ages of several hundred thousand years and even, in ideal conditions, back to about a million years. It indicates that these dunes were accumulating more or less continuously between about 16,000 and 10,000 years ago, before the lake came into being and before people moved in.

There were two main phases of prehistoric human occupation at Gobero. The first was between about 9700 and 8200 years ago and was similar to the Kiffian culture first identified at Adrar Bous. The skeletons show that the people involved were very tall, and the chemical composition of their teeth[31] indicates that they were more or less sedentary hunter-fisher-gatherers. They were equipped with pottery, bone harpoon points, and bone fish-hooks and a tool kit made up of carefully worked small pieces of sharp or pointed stones known as microliths (literally, 'small stones'). People belonging to this culture ranged widely across the southern and central Sahara. Pollen grains extracted from some of the burial sites show that the vegetation was open savanna with grasses, sedges, and some trees, including fig and tamarisk. Saltbush and plants tolerant of aridity were growing on the sand dunes. The existence of rushes revealed that permanent water and wetlands were also common at this time. The lake itself was about five metres deep, but occasionally rose to about ten metres' depth before overflowing through a gap in the dunes. Nile perch (*Lates niloticus*) up to two metres long swam in the lake, which also provided a habitat for crocodiles, hippos, and turtles as well as mussels (*Mutela*) and catfish. Analysis of the strontium content of the human teeth, which reflects both diet and local rock types, indicates that the people did not move far and seem to have been able to provide for their needs without requiring long-distance foraging or seasonal migrations.

Occupation came to an end when the lake dried out during a prolonged dry phase between 8200 and 7200 years ago, when there is no evidence of human occupation. This very arid interval is evident across the Sahara as well as in the Nile valley and was associated with very cold conditions in Greenland and the North Atlantic and a sudden weakening of the tropical summer monsoon in Africa.[32]

When climatic conditions improved once more and the lake re-filled, a new group of people moved in to occupy the area around the lake between about 7200 and 4500 years ago. They were much shorter in stature and more gracile than the earlier people. In addition to hunting and fishing for their food, they brought domestic cattle (*Bos taurus*) with them. Like those before them, these newcomers also buried their dead and left behind ornaments and stone artefacts with some of the burials. From about 4500 years ago onwards, the climate became arid once more and people abandoned the area for good once the lake became dry. As might be expected, both Gobero and Adrar Bous share a similar environmental history. The main difference is that the story that has been unravelled at Gobero is vastly more detailed and based on many more dates.

Neolithic Cattle Herders of the Sahara

By the start of 1970 sporadic finds of a few bones, teeth, or horn-cores belonging to prehistoric cattle thought to be domestic breeds, together with rock paintings depicting herds of brindled cattle and their human herders, all pointed to the widespread presence of cattle herders in the Sahara 5000 and more years ago. However, no complete skeleton had yet been found. Our chance discovery on our first morning at Adrar Bous of a white horn-core protruding from the dark clay which made it more visible led to the excavation of the complete skeleton of a short-horned domestic cow, *Bos brachyceros*, which had died in a small swamp about 5000 years ago[33] (plate 3.4). During times of drought when the expected rains are delayed by weeks or longer, both wild and domestic animals tend to converge on the muddy pools that are all that is left of their rapidly dwindling water supplies. Many of these animals become bogged and too weak to clamber from the sticky mud, and therefore perish where they collapse. Unless eaten by scavengers, some become preserved as the mud dries out and hardens and can remain buried and fossilised for thousands of years. It took us fourteen days to exhume the cow using dental picks and brushes, and great care was needed because the bones were so brittle, but it was worth the effort.

There is still a vigorous debate among archaeologists as to whether cattle were first domesticated from the herds of wild aurochs that once roamed the Sahara or were brought in from the Fertile Crescent region of the Levant and Anatolia, or whether both could have occurred.[34] Only a programme of genetic research and rigorous dating of well-preserved samples of cattle teeth and bones will supply answers to this puzzle. The few reliable ages that are available for genuinely domesticated Neolithic cattle (as opposed to hunted wild cattle) appear to show that the Neolithic sites become progressively younger as you proceed westwards from the Nile Valley. In the Nile Valley itself, the ages appear to become younger as you go south, suggesting that the early cattle herders arrived via the Sinai and moved gradually south along the Nile Valley and progressively westwards across the Sahara.[35]

At Adrar Bous the final stages of the Neolithic lake are revealed by a concentration of freshwater shells along the outermost lake shoreline. When Mr Bruce W. Sparks at the University of Cambridge studied the bulk samples of the shells I had collected, he expressed surprise at the apparent absence of land snails. He attributed this to a lack of grass around the lake to provide food for the snails. In discussion we both wondered whether the lake had been fed mainly from below by groundwater at a time when the local water table was still high. As the climate became drier, the groundwater level would start to sink, and the lake would dry out, but this would probably not be a sudden event, and there may have been a time lag of decades or even centuries to buffer the effects of regional drought. Once the lake had dried up, the cattle herders would have had little choice but to leave.

The approximate date of their departure is interesting. If it was indeed close to 4500–4200 years ago, as the dated evidence from both Gobero and Adrar Bous seems to indicate, this date coincides with a time when a sustained reduction in Nile flow may have brought about the final demise of the Old Kingdom regime in Egypt.[36] It may also be the same prolonged drought that led to the fall of the Akkadian Empire in Mesopotamia[37] and, perhaps, of the Indus Valley Culture in the Thar or Rajasthan Desert of northwest India and the Cholistan Desert of eastern Pakistan.[38]

Underground Water and Horse-Drawn Chariots

The Sahara was not entirely abandoned by humans after this time. Permanent water was still quite widely available from groundwater springs at a number of places across the desert. Later migrations by Iron Age people saw a number of Saharan oases becoming occupied or re-occupied. In the Fezzan region of the southwest Libyan Desert, the Garamantes tribe mentioned by Herodotus[39] about 2400 years ago made effective use of two innovations brought in from outside North Africa. One was the use of underground tunnels and wells to tap sub-surface groundwater and use it for irrigation on a very large scale. This system of underground water extraction using tunnels known as *qanats* in Persian or *foggara* in Arabic was introduced by the Persians when they first invaded Egypt in 525 BC under the command of King Cambyses II and is still in use today in remote parts of the Sahara. The second innovation that gave the Garamantes the edge in local warfare was the use of horse-drawn chariots (plate 3.5). The Hyksos[40] introduced the horse into Egypt during their invasion in about 1660 BC and the use of horses soon spread across the desert. There are rare paintings of horses at Jebel 'Uweinat (map 1) in the far southeast of the Libyan Desert and engravings of horses at a number of sites in the Aïr Mountains, where they are shown being led on a leash by tall, plumed warriors. For the controversial author Erich von Däniken,[41] these were the 'chariots of the gods' and the plumes, antennae for superior beings from outer space. Quite why these superior beings would have needed horses is anyone's guess.

The Romans in North Africa found the marauding Garamantes trou-blesome.[42] In 19 BC, the Roman legate Cornelius Balbus conquered the Fezzan (map 1). Pliny describes his subsequent triumph and mentions that Balbus reached the Dasibari River in the far south. The present-day Songhoi inhabitants of the middle Niger River near Timbuktu (map 1) refer to the river as Da Isa Bari, or the 'Great River of the Das'. Running diagonally across the Sahara from Tripoli (map 1) in the north to Gao (map 1) on the Niger are rock engravings and paintings of horse-drawn chariots. In his account of the Tassili frescoes, Henri Lhote[43] suggests that charioteers may have crossed the Sahara and reached the Niger by about 1000 BC, but this remains conjectural.

Pollen from Crocodile Coprolites
and Less Exotic Sources

In addition to the rock art, which is not easy to date accurately, there are many other bits of evidence we can use to reconstruct past environments in the Sahara. One such piece of evidence comes from the humble pollen grain, which shows what the past vegetation was like. Pollen grains have a very tough external protective cover, and under suitable conditions they can survive as fossils for thousands or even millions of years, especially if deposited in swamps or wetlands. They fare less well in deserts, where blowing sand can destroy them or where extreme aridity causes them to disintegrate.

People who devote their professional lives to studying fossil pollen grains are remarkable beings. In addition to needing immense patience when counting hundreds of pollen grains under the microscope, and great care and courage in using some extremely nasty chemicals to extract pollen grains from the sediments in which they were preserved, they also need a good measure of luck in finding suitable samples.

During their brief visit to Adrar Bous in November 1959, members of the first French Berliet expedition driving down from the oasis of Djanet (map 1) in southern Algeria across the Ténéré Desert to Chad had observed some curiously shaped stones composed of calcium carbonate or limestone. Some of these carbonate concretions were very likely to have been fossilised crocodile dung pellets or, more poetically, 'calcified crocodile coprolites'. The geologist Hugues Faure had also worked at Adrar Bous when mapping the geological formations east of the Aïr Massif and was able to obtain a radiocarbon age for one such sample which showed that the lake at Adrar Bous had dried out more than 8000 years ago. Similar crocodile coprolite fossils had been recovered from former lake sediments in the Hoggar Mountains of southern Algeria and were found to contain pollen grains blown in from the plants that were growing on the mountain slopes at that time.[44] Pollen from less exotic samples such as now dry lakebeds and wetlands also proved revealing. For example, pollen recovered from three sites in hyper-arid northern Sudan indicate that the climate was wetter between about

10,000 and 5000 years ago, consistent with a northward displacement of the summer rainfall zone by about 400–450 kilometres.[45]

What was interesting from the pollen analysis of samples from upland sites in the Sahara such as the volcanic highlands of the Hoggar, Tibesti, and Jebel Marra (map 1) was that it indicated vegetation consisting of both Mediterranean plants and tropical plants, suggesting that it had once been possible for Mediterranean plants to migrate south across the Sahara and for tropical savanna plants to migrate north into the central Sahara.[46]

It soon became clear that there had been multiple migrations of plants across the Sahara from just about every possible direction and that there had been multiple phases of prolonged aridity interspersed with wetter climatic intervals. These plant migrations were not simply into the Sahara but also from the Sahara out across Eurasia. If you visit the Rajasthan Desert[47] in northwest India, you will very soon spot dozens of trees and shrubs and grasses that are familiar to the people living along the southern Sahara, located more than 5500 miles or 9000 kilometres to the west. Another intriguing discovery is that plants adapted to aridity were already present in the mountains of the Sahara well before this time. In fact, there was a very long interval of adaptation to a dry environment dating back to seven million years ago, which is when the Sahara first became dry. I discuss this in chapter 4.

Saharan Rock Art

Because deserts are so dry, they operate as gigantic natural museums and allow even delicate rock paintings and rock engravings to survive for many thousands of years after they were first made.[48] Among the animals painted or engraved on the rock by prehistoric artists in the Sahara are giraffes (plate 3.6), elephants (plate 3.7), rhinos, and many types of antelope (plate 3.8). Needless to say, such animals would not find much to eat or drink in the present-day Sahara. The more recent rock art dating from Neolithic times about 5000 years ago shows scenes of cattle camps, mothers and babies, men armed with bows and arrows and often accompanied by dogs; some are engaged in herding their cattle, others are engaged in hunting or even raiding. One remarkable scene from the Tassili sandstone plateau in central Algeria shows women riding on the backs of oxen (plate 3.9) and sheltered beneath tent-like structures similar to those used today by the Baggara cattle nomads of western Sudan during their annual summer migrations north in search of fresh pasture.

The French amateur archaeologist Henri Lhote spent many seasons tracing what became known as the 'Tassili Frescoes'. I met Henri Lhote in early January 1970 in the former French Foreign Legion fort since refurbished as the Hotel de l'Aïr in Agadès. Conversation proved somewhat tricky because by then he was almost completely deaf. His tracings are stored today in a highly secure underground room in the Museé de l'Homme in Paris which I visited on 1 June 2018. The museum staff members mentioned that the tracings on occasion reveal a measure of artistic license rather than what some might consider unduly slavish scientific accuracy!

Another remarkable scene from the Tassili shows a canoe identical in shape and form to the *tankwa* canoes used to this day by fishermen on Lake Tana in Ethiopia. These one-person canoes are made from papyrus reeds. We can conclude that watercraft were in use in the heart of the Sahara at a time when papyrus reeds were growing in local swamps and wetlands that have long since vanished from this area.

Some of the most vivid prehistoric rock paintings are to be found at Jebel 'Uweinat, an isolated mountain located in what is now one of the driest parts of the Sahara, athwart the frontiers of Libya, Egypt, and Sudan. Its name in Arabic means mountain of small springs and even today it still has several permanent sources of fresh water.[49] More than a thousand individual rock paintings have been recorded at Jebel 'Uweinat, which is about 25 kilometres wide and rises to 1934 metres in elevation. The western side of this ring-complex is deeply dissected by two major wadis, and many of the paintings are on suitable smooth rock outcrops along the sides of these two desert mountain valleys. There are many generations of rock paintings, and many of the younger paintings are superimposed upon older paintings so that a relative chronology of different styles can be established.

Because it is so difficult to date rock paintings directly, archaeologists use other evidence that they have collected and dated from their excavations as a way of placing some of the paintings into a plausible time sequence. I will illustrate this using the evidence obtained by archaeologists from the Gilf Kebir (map 3) in the Western Desert of Egypt to place the rock art at Jebel 'Uweinat and Jebel Arkenu (map 1) in the far southeast of the Libyan Desert into a prehistoric time context. Using pottery styles and prehistoric stone tool assemblages, German archaeologists have identified four main phases of prehistoric occupation in the Gilf Kebir.[50] The oldest phase (Gilf A) is a hunter-gatherer phase dating back to about 10,300 years ago. It was followed by the Gilf B phase (8800–6300 years ago) which was one of hunter-gatherers who made pottery. The Gilf C phase (6300–5300 years ago) was one of Neolithic herders of sheep, goats, and cattle. There was growing aridity during the Gilf D phase (5300–4700 years ago).

In the absence of any other suitable evidence, rock art specialists working at Jebel 'Uweinat have drawn upon dated occupation sites in the Gilf Kebir and elsewhere to establish a provisional occupation chronology for four occupation phases at 'Uweinat. Phase A is without pottery and is probably older than about 8600 years. It may coincide in part with a wetter phase at 'Uweinat dated between 9400 and 8100 years ago.[51] Another phase of hunter-gatherer occupation, this time with pottery, has a likely age range between about 8800 and 6300 years ago. By analogy with the Gilf Kebir occupation sites, the main period of Neolithic occupation at 'Uweinat associated with paintings showing cattle, sheep, goats, dogs, and archers, as well as domestic scenes with men, women, and children, has a likely age range between about 6300 and about 5300 years ago.

Many of the paintings probably had religious significance, just as they do today among the San people of the Kalahari and the Australian Aborigines, and they are not to be interpreted too literally (plate 3.10). For example, in an elevated sandstone basin surrounded by rugged sandstone cliffs at Jebel Arkenu, a small mountain lying forty kilometres northwest of 'Uweinat, there is a painted rock shelter showing many giraffes. It would be pretty impossible for giraffes to reach this spot; they were most likely browsing out on the plains surrounding the mountain.[52] The artists were therefore portraying in one locality what they had observed in another, more distant locality. It is also possible that the giraffes depicted had some spiritual or ceremonial significance, which is the case today among the San.

Wet or Dry?

One is tempted to ask whether the Sahara ever ceased to be a desert and whether the scattered finds of prehistoric human occupation and rock art in different areas of the Sahara simply denote a few favoured localities akin to the desert oases of the present day. The answer comes from many independent lines of evidence (see table 3.1) and points to a number of occasions when the Sahara was criss-crossed by active rivers and studded with an abundance of freshwater lakes, some huge, others quite small. The most recent of these wetter climatic intervals is bracketed between roughly 15,000 and 5000 years ago, although it was not continuously wet, as I have discussed earlier in this chapter. Indeed, the evidence for widespread use of water resources by prehistoric groups in the southern Sahara at this time so impressed the British archaeologist John Sutton[53] that he referred to this phase as the African *aquatic* cultural phase and the tool kit as *aqualithic*. These terms are of course something of an exaggeration. However, they do signal that in geologically recent times, the Sahara was for many thousands of years a green and pleasant land throughout most if not all of its present extent. Occasional dry phases interrupted these wetter times, but for much of the time until about 5000 years ago, life in the Sahara would have been congenial for plants, animals, and human societies that would find life there today tough if not impossible. What caused the change is the subject of part two.

PART TWO

A SEA OF SAND

Part two deals with the progressive desiccation of the Sahara and with the birth and spread of the great sand seas and dune fields that now cover a fifth of the Sahara. The onset of aridity was gradual and was often interrupted by wetter climatic intervals. The causes of these alternations from wetter to drier climates are linked to changes in the path of the earth around the sun and to changes in the tilt of the earth's axis, so that long-term climatic fluctuations in the Sahara are closely bound up with global climatic changes, all of which had a profound influence upon plants, animals, and prehistoric human societies.

Chapter 4 asks why the Sahara is currently so dry and rejects the notion that humans were the cause of its aridity, pointing out that the Sahara existed as a desert millions of years before humans ever appeared on the scene. The onset of Saharan aridity was associated with the slow northward drift of what is now the Sahara into dry tropical latitudes. The drying up of the former Tethys Sea, of which the present Mediterranean Sea is but a shrunken remnant, also deprived the Sahara of a major source of moist air from the northeast.

In chapter 5 we consider how the great sand seas of the Sahara came into being and note that there is a constant battle between wind and water, with desert rivers jockeying for supremacy and achieving it during times when the climate was wetter, before becoming buried beneath advancing sand dunes during times when sustained aridity led to reactivation of the previously vegetated and stable dunes.

Chapter 6 looks at the sometimes surprisingly beneficial role of Saharan desert dust in far-flung parts of the world and also shows how dust deposits can be recycled to form potentially fertile alluvial deposits. It also describes how extreme sand and dust storms can be quite lethal, drawing upon anecdotes retailed to the Greek historian Herodotus during his visit to Egypt some 2500 years ago.

Chapter 7 introduces the earliest humans to move into and occupy the desert during the relatively brief intervals when the Sahara enjoyed a wetter climate. The first human arrivals came equipped with fire and a modest stone tool kit. Later human migrants brought with them an increasingly specialised set of stone tools and adopted a lifestyle based on gathering plant foods, opportunistic hunting of the animals that roamed the Saharan savanna woodlands and grasslands, and fishing the lakes and rivers that criss-crossed the Sahara during the intervals of more prolonged and more widespread summer monsoons from the south and more sustained incursions of winter rains from the north. About 7000 years ago the hunter-gatherer-fisher societies were replaced by pastoral groups who brought herds of cattle, sheep, and goats into favourable localities in the Sahara, where there was both adequate grazing and reliable supplies of permanent fresh water. With a return once more to a more arid climate from about 5000 years ago onwards, these pastoral groups moved out of the Sahara into the well-watered regions of West Africa along great river valleys like the Tilemsi valley in Mali or into the Chad Basin with its fluctuating but permanent lake or eastwards to the Nile Valley, where perennial water could be guaranteed.

CHAPTER 4

Through a Glass Darkly

The houses are all built of salt-blocks—an indication that there is no rain in this part of Libya; for if there were, salt walls would collapse.

HERODOTUS, *THE HISTORIES*

In this chapter we try to answer two important questions relating to the origin of the Sahara, namely, when did it become dry and why? First, it will be useful to consider why the Sahara is so dry today and why it receives so little rain.

Why Is the Sahara Dry?

On page 202 of their popular book *Population, Resources, Environment: Issues in Human Ecology* published in 1970, the distinguished environmental scientists Paul and Anne Ehrlich wrote as follows:[1] 'The vast Sahara desert itself is in part man made, the result of overgrazing, faulty irrigation, and deforestation, combined with natural climatic changes. Today the Sahara is advancing southward on a broad front at a rate of several miles a year'.

This statement is both factually incorrect and misleading. We will explore this important issue in appropriate detail in chapter 8. Here it is enough to say that the Sahara was in existence as a desert millions of years before ancestral humans first began to fashion stone tools some 2.5 million years ago in the Gona Valley (see chapter 7, fig. 7.1) of the

Ethiopian Afar Rift.[2] Furthermore, we have long known that during prolonged droughts the plant cover dies off or becomes dormant along both margins of the desert, giving the illusion that the desert is expanding, while the reverse occurs during prolonged wet phases. One drought does not a desert make!

The present-day Sahara is dry for sound and immutable geographical reasons that have nothing to do with any human impact. The Sahara today is in the wrong latitude to be anything but a desert. Let me explain what I mean by this rather cryptic statement. For much of the year the sun is directly overhead at or close to the equator, so that the land and the sea in these latitudes absorb the short-wave radiation from the sun, become warm, and heat the air above them. This warm air then rises by convection to heights of 12–15 kilometres above the earth's surface. As air rises, it expands and cools. Water vapour condenses as the air cools, and intense convectional downpours occur, mostly in mid-afternoon, following prolonged daytime warming and the progressive buildup of often very high convectional clouds. Because of the tilt of the earth's axis, the zone of maximum solar heating migrates north in the northern summer and south in the southern summer over a distance of about twenty-three degrees of latitude.

This seasonal displacement of the tropical rain-bearing winds is known as the Hadley Circulation (fig. 4.1). It was in fact first described by Sir Edmund Halley (of Halley's Comet fame) in 1686 and described in greater detail fifty years later, in 1735, by Sir George Hadley.[3] Once the cool air masses aloft have shed their moisture, they begin to subside. This subsidence occurs in about latitude 23 degrees north or south, depending on the season (fig. 4.2). As the air subsides it becomes compressed and starts to heat up, just as the compressed air inside a bicycle tyre becomes hot as we pump it up. Hot air can absorb far more water vapour than cold air, and this is reflected in the aridity prevalent in these latitudes. In short, the main reason that the Sahara is so dry today is that it is located in tropical latitudes characterised by dry subsiding air masses which act as gigantic desiccators.

A second cause of Saharan aridity is the sheer size of the Sahara. It is our largest tropical desert. The distance from the Atlantic coast of

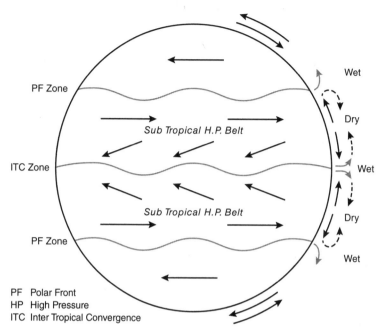

PF Zone

Sub Tropical H.P. Belt

Wet

Dry

ITC Zone

Wet

Sub Tropical H.P. Belt

Dry

PF Zone

Wet

PF Polar Front
HP High Pressure
ITC Inter Tropical Convergence

FIGURE 4.1. Global atmospheric circulation, showing location of the Hadley cells.
After M. Williams (2014). *Climate Change in Deserts: Past, Present and Future.* Cambridge
University Press, Cambridge and New York, fig. 2.1. © Cambridge University Press.
Reproduced with permission of the Licensor through PLSclear.

Mauritania to the arid Red Sea Hills in the east is 4800 kilometres. From
south to north the Sahara is never less than 2000 kilometres wide. As a
result, moist air masses blowing in from the Atlantic in the west and
south, the Indian Ocean in the east and southeast, and the Mediterranean in the north very soon shed most of their moisture and so bring
very little precipitation to the Sahara.

High mountains also have an influence. The Atlas Mountains in the
northwest and the Ethiopian and East African Highlands in the east
receive abundant precipitation in the form of rain or even snow. This is
because as moist air rises over the mountains, the air expands and becomes cooler, so that the water vapour in the air soon reaches its satura-
tion point and surplus moisture is released as precipitation over the

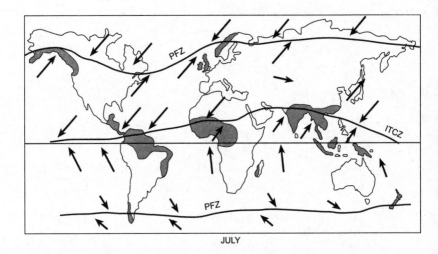

JULY

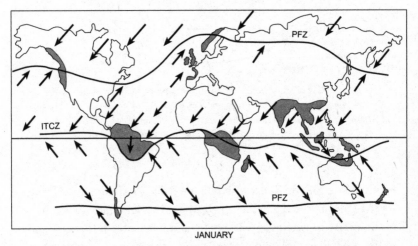

JANUARY

FIGURE 4.2. Seasonal migration of global wind systems and of the Intertropical Convergence Zone (ITCZ). PFZ is the Polar Front Zone. After M. Williams (2014). *Climate Change in Deserts: Past, Present and Future*. Cambridge University Press, Cambridge and New York, fig. 2.2. © Cambridge University Press.
Reproduced with permission of the Licensor through PLSclear.

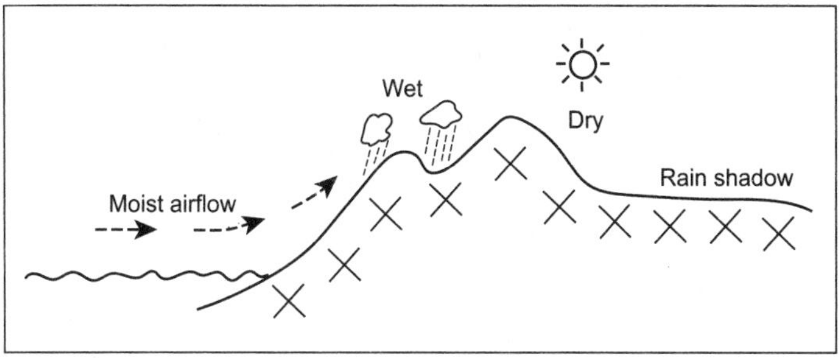

FIGURE 4.3. The rain-shadow effect.

highlands. As the air masses pass over the mountains and reach the lowlands (in this case the Sahara), they have lost much of their initial moisture content, so that the lowlands to leeward of hills and mountain are described as being in the rain-shadow of the uplands (fig. 4.3).

Another factor contributing to the present-day aridity of the Sahara is the presence of a cold ocean current flowing southwards along the west coast of the Sahara. Wind blowing across the cold ocean surface onto the land will become somewhat warmer and so will be less likely to lose moisture in the form of rain. During times when the Trade Winds are strong, especially during glacial periods when the temperature and atmospheric pressure gradients between the equator and the pole are much steeper, cold water from deep in the Atlantic Ocean comes to the surface and enhances the influence of the cold ocean current offshore. This accentuates aridity on land (fig. 4.4).

Drifting into Drier Latitudes

We can now refine our opening question. How and when did the Sahara reach its present latitude? In order to answer this question, we must first look back to the time more than 400 million years ago when Africa formed part of a super-continent called Gondwana, which consisted of Africa, South America, Antarctica, Australia, and India (see fig. 1.1). By about 200 million years ago, Gondwana had begun to split into the two huge continents of West Gondwana (Africa and South America) and

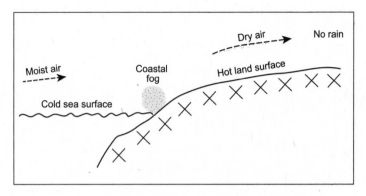

FIGURE 4.4. Cold water offshore accentuates aridity on land.

East Gondwana (Australia, Antarctica, and India). This separation was completed during the Jurassic, roughly 200–145 million years ago, a period well known for its dinosaurs. Table 1.1 in chapter 1 lists the names and ages of the geological periods mentioned in this chapter.

Continued breakup of these two very large continents occurred during the Cretaceous period roughly 145–66 million years ago, at the end of which the dinosaurs became extinct. During this time, about 130 million years ago, Africa separated from South America and the Atlantic Ocean began to develop between these two continents and between Africa and North America (see fig. 1.2), which had been part of another super-continent called Laurasia.

During much of the Cretaceous, Africa lay well south of its present location and was aligned further to the west. As a result, the equator ran diagonally across what is now the Sahara from southern Nigeria through central Chad, northern Sudan, and Egypt into Arabia.[4] As Africa moved northwards into latitudes characterised by dry subsiding air during the late Cretaceous and the following Cenozoic era (66 million years ago to present), the equatorial rainforest which had once covered the Sahara was displaced southwards and desiccation gradually set in.[5] As the northward drift of Africa was at the fingernail growth rate of only a few centimetres a year, all of these events took place very slowly and there was plenty of time for plants and animals to adapt to the progressive changes in latitude or to migrate.

Further to the east, uplift of the Ethiopian highlands was heralded by a massive outpouring of very fluid volcanic rocks known as the Trap Series. Most of these basaltic lavas were erupted during quite a short interval of geologic time centred on 30 million years ago.[6] As the Ethiopian highlands were being uplifted, possibly as a result of their location over a source of very hot molten rock, in neighbouring South Sudan deep rift valleys became filled with up to 15 kilometres of sediments, some of them highly organic. These are the parent rocks of the oil-bearing rock formations in that region today. Some of the initial rifting and sedimentation within the rifts may be much older, perhaps dating back to Mesozoic times about 250–66 million years ago.[7]

Continuing uplift in Ethiopia and East Africa during the past 6–8 million years created a large topographic barrier which caused a change in atmospheric circulation, leading to reduced rainfall in East Africa and the Chad Basin.[8] One result of this change in rainfall regime over eastern Africa, including Ethiopia, was a change from tropical forest to open grassland and woodland. This is also the time when we see the appearance of a series of hominins that are unique to Africa (see chapter 7).

An additional line of evidence consistent with the slow northward drift of Africa is one that I discussed in chapter 2. In northern Nigeria and southern Niger there are a series of curious geological features called ring-complexes. These ring-complexes are aligned very roughly from south to north. Ages obtained for these features show that they become progressively younger to the south and older to the north, indicating that they may have developed when Africa drifted northward over a very hot zone deep in the earth's crust, as illustrated in figure 2.4.

First Signs of Aridity

There was a slight clockwise rotation of the African plate during the Miocene (23–5.3 million years ago) and Pliocene (5.3–2.6 million years ago) which brought it into contact with Eurasia.[9] The collision led to renewed uplift of the Atlas Mountains in northwest Africa and happened at approximately the same time as a phase of widespread volcanic activity and uplift in the Sahara, which gave rise to the Hoggar, Tibesti, Aïr, and Jebel Marra volcanic uplands much as they are today (see chapter 2, fig. 2.5). Both the Hoggar and the Aïr consist of very much older rocks, with the latest episode of volcanic activity merely adding the volcanic icing to the preexisting cake. This was a time of widespread crustal deformation and associated faulting and folding across North Africa. The major elements of Saharan topography were created at this time, including the pattern of swells and depressions noted by the geologist Arthur Holmes in the 1940s and dated more precisely since then.[10] As we saw in chapter 2, in the northern Aïr Mountains, Cretaceous marine sedimentary rocks occur at an elevation close to 2000 metres above sea level as a result of faulting and uplift. East of the mountains these Cretaceous marine rocks have been downfaulted and now occupy buried rift valleys.

Three other influences contributed to the gradual drying out of the Sahara. Some of them may at first seem quite surprising. They are uplift of the Tibetan plateau, accumulation of ice over Antarctica and in high latitudes of the Northern Hemisphere, and cooling of the world's oceans.

Uplift of the vast Tibetan plateau in the late Miocene led to the intensification of the easterly jet stream that now brings dry subsiding air to the deserts of Pakistan, Afghanistan, Iran, Arabia, Somalia, and North Africa.[11] There was also a significant change in the late Miocene flora and fauna of East Africa and the Potwar Plateau in the Himalayan foothills of Pakistan, indicating a major change in climate at that time and a trend towards a more seasonal climatic regime with a long dry season.[12] Marine sediment cores recovered to the west of the Sahara also suggest drier conditions at this time, evident in an increase in the input of wind-blown Saharan desert dust to the Atlantic.[13] Saharan dust also occurs in soils of Pliocene age in the Canary Islands off the northwest Saharan coast.[14]

It may be hard to imagine, but the progressive accumulation of ice in Antarctica[15] also played a part in the drying out of the Sahara. Mountain glaciers were present in Antarctica early in the Oligocene, and a large ice cap had formed by at least 10 million years ago. Accumulation of ice in the Northern Hemisphere began much later, and only really took off about 2.5 million years ago. As the poles became colder, high-latitude sea-surface temperatures also became colder. As a consequence, the temperature and atmospheric pressure gradients between equator and poles also increased. Trade Wind velocities increased as a result of the steepened pressure gradient, as did their ability to mobilise and transport Saharan alluvial sands and sculpt them into desert dunes.

At present, about two-thirds of our global precipitation falls between latitudes 40°N and 40°S and is controlled primarily by evaporation from warm tropical seas.[16] Miocene cooling of the ocean surface[17] would also have resulted in reduced precipitation over the Sahara. Table 4.1 shows the key factors associated with the progressive desiccation of the Sahara.

From Rain Forest to Desert Scrub

During most (but not all) of the Palaeocene and Eocene, about 66–34 million years ago, much of what is now the southern and central Sahara enjoyed a hot, wet climate and was covered in equatorial rain forest.[18] Chemical weathering was intense beneath the highly organic and densely vegetated soil surface, and rocks exposed near the surface were weathered to depths up to fifty metres,[19] a process known as deep weathering. During the ensuing Miocene (23–5.3 million years ago), this region experienced uplift as well as climatic desiccation.[20] The climatic desiccation already under way was enhanced by two independent factors. One was the late Miocene shrinking of the vast Tethys Sea[21] as Africa moved north towards Eurasia. The Mediterranean Sea is the shrunken remnant of this once extensive sea. As a consequence, northern Africa was deprived of an abundant supply of moist air blowing from the Tethys Sea. The second critical factor was late Miocene global cooling[22] about 8–6 million years ago which saw the spread of the modern terrestrial plant and animal ecosystems so familiar to us today.

Differential Miocene uplift across the Sahara combined with climatic desiccation and a reduced protective plant cover ushered in a period of

TABLE 4.1. Key factors associated with the progressive desiccation of the Sahara and some of the outcomes

50–45 Ma	Separation of Australia from Antarctica; inception of Southern Ocean.
45 Ma	Collision of Greater India with Asia. Progressive uplift of the Tibetan Plateau and Himalayas.
34–33 Ma	Opening of the Drake Passage between Antarctica and South America. Creation of circum-Antarctic Current. Cooling of the Southern Ocean. Ice cap growth in Antarctica (34 Ma). Late Miocene ice sheet in West Antarctica. Major global cooling. Severe desiccation in central Asia. East Africa: uplift and rifting (~ 25 Ma). Africa meets Europe: mountain building (15 Ma).
7 Ma	Tethys Sea shrinks: Rainfall less and more seasonal. Global cooling and desiccation. Savanna woodland and grassland replace forest in Africa and Asia.
6–5 Ma	Miocene salinity crisis: Mediterranean salt desert. Incision of Nile canyon. Genetic isolation of Africa from Eurasia (5.96–5.33 Ma). Emergence of bipedal hominins in the Afar Rift of Ethiopia.
4–3 Ma	Closure of the Indonesian Seaway. Diversion of cool ocean water towards East Africa. Desiccation in East Africa. Closure of the Panama Isthmus.
2.7–2.5 Ma	Rapid accumulation of ice over North America. Cooling and tropical desiccation (2.6 Ma). Drying out of East Africa and the Sahara. First appearance of stone toolmaking in East Africa.
2.4–0.9 Ma	High-frequency, low-amplitude 41,000-year glacial-interglacial cycles.
0.9–0 Ma	Low-frequency, high-amplitude 100,000-year glacial-interglacial cycles.

Note: Ma are millions of years before present.
Source: Adapted from M. Williams (2014). *Climate Change in Deserts: Past, Present and Future*. Cambridge University Press, Cambridge and New York, pp. 24–28.

intense erosion of the deep weathering mantle and exposure of the irregular weathering front[23] (see chapter 2, fig. 2.6). In many parts of the Sahara today the geomorphic legacy of this process consists of large boulders perched somewhat precariously on the rocks beneath. Uplift of the Saharan uplands caused a wave of fluvial erosion. Rivers flowing from the uplands deposited gravels, sands, and clays during the final phase of widespread fluvial erosion across the Sahara.

The origin of the Sahara as a desert probably stems from about this time. However, the Sahara was not as universally arid during the Miocene as it is today. There were quite long intervals of wetter climate[24] during which Mediterranean plants migrated slowly southwards into the central and southern Sahara, while plants from the wet tropics moved slowly northwards. Today the only evidence of these great migrations consists of some sparse relict populations of both Mediterranean and tropical plants that now occupy mountain refugia in the Hoggar, Tibesti, Aïr, and Jebel Marra.[25] Plant migration at these times was facilitated by the presence of former waterways and probably took place along their valleys. Some very big rivers flowed northwards from what is now the Chad Basin across the Sahara to the Mediterranean.[26] These rivers carved very wide valleys, some of which are clearly visible today on the ground in the southern Libyan Desert between Tibesti volcano and the three large sandstone plateaux located just to the east of the northern tip of Tibesti.[27] They are even more spectacular when seen from space.[28] The combined evidence from plant and animal fossils and sediment geochemistry also confirms that the climate in the headwaters of these Miocene rivers was relatively humid, with the late Miocene vegetation in the Chad Basin comprised of a mosaic of wetlands, savanna grasslands, and woodlands.[29] But a very strange surprise was in store for the Mediterranean coastal regions of North Africa.

The Mediterranean Becomes a Salt Desert

A very remarkable series of events took place during the late Miocene between 5.96 and 5.33 million years ago. At intervals during that time the Mediterranean was cut off from the Atlantic, dried out within a few centuries each time, and became a salt desert.[30] Some geologists have suggested that the repetitive drying out of the Mediterranean was caused by earth movements. I do not find this suggestion convincing. I think it is far more likely that we are dealing with global fluctuations in sea level linked to the growth and decay of ice caps in West Antarctica.

During glacial intervals, as the ice caps grew, global sea level fell, until the shallow sill at the western end of the Mediterranean located beneath the present Straits of Gibraltar emerged above sea level and acted as a very effective dam, preventing the free flow of Atlantic surface water into the Mediterranean Basin and outflow at depth from the Mediterranean. The sill now has an average depth of about 365 metres, and at the point where Africa is closest to Europe, it is about 300 metres deep. It must have been quite shallow in the late Miocene and would have become deeper since then as a result of marine erosion. In step with the glacial intervals, phases of inflow when ice volume was low and sea level was high alternated with phases of desiccation. The end result was accumulation of a layer of salt and other evaporites up to about one kilometre thick on the floor of the Mediterranean.

Whatever the cause, desiccation of the Mediterranean would have deprived northern Africa of a major source of moisture. These salt deposits are now buried beneath a protective layer of marine Pliocene sediments. A similar salt desert formed at this time on the floor of the Red Sea so that Africa was probably isolated genetically from Eurasia for much of that time, perhaps facilitating the emergence of our hominin ancestors in Africa and nowhere else.

Another consequence of the repeated drying out of the Mediterranean was repeated lowering of the outlets of all rivers draining into the Mediterranean, leading to the formation of some very deep river canyons. The Nile, for example, cut down 2.5 kilometres north of Cairo and flowed at 170 metres below present sea level at Aswan, located 1200 kilometres upstream of the present Nile delta.[31] During that time, the Nile eroded about 80,000 cubic kilometres of rock from its canyon and deposited it on the floor of the eastern Mediterranean.

The net effect of all the factors listed in table 4.1 was the gradual emergence of the vast and arid land that we today refer to as the Sahara. The stage was now set for the birth and growth of desert dunes and the great sand seas that now cover about one-fifth of the Sahara. Chapter 5 tells this tale of wind, water, and sand.

CHAPTER 5

Water and Sand

At times, especially on a still evening after a windy day, the dunes emit, suddenly, spontaneously, and for many minutes, a low-pitched sound so penetrating that normal speech can be heard only with difficulty.

R.A. BAGNOLD, *THE PHYSICS OF BLOWN SAND AND DESERT DUNES*

Sand Dunes and Sand Grains

The mystical English poet and painter William Blake (1757–1827) urged us in his poem *Auguries of Innocence* (?1803) to use our imagination to *see a world in a grain of sand*. I will try to follow his advice. When we think of the Sahara, we think of sand; when we think of sand, we think of sand dunes. Where does all the sand come from? Although only about one-fifth of the Sahara is actually covered in sand dunes and sandplains—the rest consists mostly of mountains, gravel plains, and vast sandstone or limestone plateaux—these dunes have exerted an almost hypnotic attraction for desert explorers.

If there is one name above all that we associate with the scientific study of Saharan sand dunes, it is that of Ralph Alger Bagnold (1896–1990).[1] The book he wrote entitled *The Physics of Blown Sand and Desert Dunes* remains a classic and forms the basis for understanding sand movement and dune formation today. He wrote it in 1939; it was published in 1941 and reprinted in 1965. He also wrote two other highly

readable books: *Libyan Sands—Travel in a Dead World* (1935) describing his explorations in the Libyan Desert between the two World Wars,
and *Sand, Wind, and War—Memoirs of a Desert Explorer* (1990). He was
encouraged to write this memoir by C. Vance Haynes from Tucson,
Arizona, who himself carried out distinguished pioneering studies of
climatic changes in the eastern Sahara after World War II, often revisit-
ing the localities surveyed by Bagnold before the war.

For much of his professional life, Bagnold was a British soldier. During World War I he was a humble sapper in the Royal Engineers and
spent much of his time defusing enemy bombs that had failed to deto-
nate. This was a game of double and triple bluff where a single error
meant death. That he survived is a testament to his calm temperament
and highly focussed analytical mind. As a reward for his work defusing
bombs, the British Army granted him leave to study engineering at the
University of Cambridge during 1919–1921. He found the degree too
theoretical to be of much practical value, but it did give him a solid
grounding in maths and physics, which he used to good effect later in
his studies of sand particle movement. During World War II Bagnold
served in the Western Desert and was the founder of the 'Long Range
Desert Group'. The men of the Long Range Desert Group used to navigate for weeks deep in the desert behind enemy lines to bring back valuable intelligence and they also helped to transport the Special Air Ser-
vice troopers to their target destinations until they acquired enough of
their own transport. The success of the Long Range Desert Group in
navigating across the trackless desert was in no small part due to the
wealth of hard-won experience in desert vehicle navigation Bagnold had
gained before the war. The sun compass that he designed was still invaluable for Saharan desert navigation in the 1960s and was only super-
seded when global positioning satellites (GPS) took the fun out of navigation but made desert travel much safer.

Using a variety of vehicles suitably modified for desert travel, Bagnold and his companions explored the Western Desert of Egypt and the
arid northern fringes of the Sudan in a series of expeditions in 1927, 1928,
1929, and 1930. The primary aim was to discover a way through the Great
Sand Sea (fig. 5.1) with its hundred-metres-high sand dunes that many

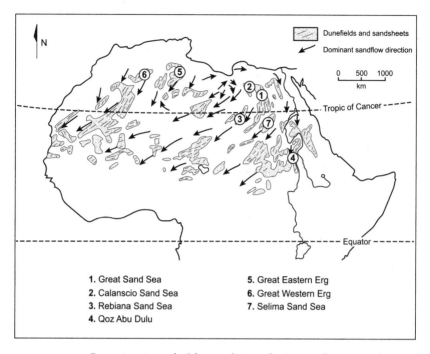

FIGURE 5.1. Dune orientation in the Sahara in relation to dominant sand-moving winds. Modified from M.A.J. Williams (1984). Geology. In *Key Environments: Sahara Desert*, J.L. Cloudsley-Thompson (ed.). Oxford, Pergamon Press, pp. 31–39, fig. 3.7. Reprinted with permission from Elsevier.

believed could not be crossed using vehicles. Impossible and impass-able. He proved them wrong, in the process discovering the hyper-arid Selima Sand Sea (fig. 5.1) and collecting useful topographic informa-tion as well as pinpointing places where water could be found. In 1932, Bagnold organised an even longer expedition of 3700 miles (6000 kilo-metres) via Jebel ʻUweinat (map 1) to the eastern slopes of Tibesti vol-cano (map 1) in Chad, south as far as El Fasher (map 1) in western Sudan, and back northeast to Cairo via the oases of Selima and Dakhla (map 3). To his abiding credit, Bagnold always took great care to bring along on his expeditions a team that included botanists, geologists, and archaeologists, and on return they published their scientific observations very promptly in a series of still valuable articles in the *Geographical Journal* of the Royal Geographical Society of London.

By the mid-1930s, Bagnold had become intrigued (perhaps obsessed is a more accurate term) by two key questions. How did sand dunes acquire and retain their very distinctive shapes? Under what precise conditions were grains of sand entrained and transported? He had seen enough dunes to know that certain repetitive dune patterns and shapes were always apparent on his desert travels. He had also experienced the affliction of a sandstorm in which only the heads of his team members were visible above the swirling sands that enveloped the rest of their bodies as they tried to walk across a sandplain. He was by now convinced that the answers to these two questions would come from experimental physics and not from purely geomorphological observations. This is borne out by what he wrote in the introduction to his 1941 book on blown sand and desert dunes: 'It seemed to me, however, that the subject of sand movement lies far more in the realm of physics than of geomorphology; and if any advance were to be made in our knowledge of it, it must in the first instance be approached via the study of the behaviour of a single grain in a stream of wind.'

His attraction to the stark simplicity of desert dune patterns and shapes is also apparent in this same introduction. 'Here, instead of finding chaos and disorder, the observer never fails to be impressed at a simplicity of form, an exactitude of repetition and a geometric order unknown in nature on a scale larger than that of crystalline structure.'

Where Does the Sand Come From?

Without sand, there would be no sand dunes, so where does the sand come from? The answer is very simple: rivers. Desert rivers are the source of desert sand. Rivers have been eroding the Saharan uplands and transporting the eroded sediment down to the desert lowlands or even, during much wetter climatic intervals, as far as the Atlantic Ocean and the Mediterranean Sea. These rivers have been active in their work of denudation (fig. 5.2) across North Africa well before the Sahara first became a desert about seven million years ago.

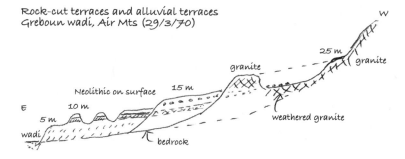

FIGURE 5.2. Sketch of rock-cut terraces and alluvial terraces in a valley near the plateau of Greboun in the northern Aïr Mountains of Niger, south-central Sahara (29 March 1970). The terraces indicate alternating phases of incision and sedimentation in the valley.

Canadian petroleum geologist Dave Griffin spent many years piecing together the history of the great Miocene rivers which flowed from the northern rim of the Chad Basin across the Libyan Desert to the Mediterranean Sea. He named them the Sahabi Rivers[2] from the locality in Chad where they originated. His evidence consisted of satellite imagery and bore log data, the latter allowing him to reconstruct the likely flow regime and climatic conditions under which the rivers once flowed. As the climate became progressively drier, and the rivers no longer flowed to the sea, they deposited their load of sand and gravel across what is now the Sahara Desert.

Other great Saharan rivers had a similar history. The Saoura River[3] (map 1) flows south from the Atlas Mountains (map 1) in Algeria close to the modern border with Morocco (map 2). To this day, during times of exceptional rainfall in the mountains, the Saoura will flow for many hundreds of miles into the desert, leaving behind its customary load of sand and gravel. The Irharhar[4] (map 1) is another such river. It flows south from the Hoggar Mountains and is thought to have crossed much of the Sahara as recently as 120,000 years ago, allowing safe passage for humans across the desert, just as the Nile does today in the eastern Sahara. A series of rivers now buried beneath the sands of the Libyan Desert and known collectively as the Kufra Rivers[5]—after Kufra Oasis (map 3)—are also thought to have flowed from the Tibesti uplands across the desert to the

Mediterranean Sea. Because so much of the evidence is fragmentary and mostly concealed beneath wind-blown sands, it is hard to reconstruct the former drainage networks with the accuracy we would like. Nevertheless, geologists are seldom deterred by too little evidence; they know that improved techniques and perseverance will add future pieces to the jigsaw.

Over time, many Saharan rivers large and small ferried their sediment loads of gravel, sand, silt, and clay down to the desert lowlands and to the vast internal drainage depressions that had been created by earth movements associated with the episodic uplift of the Saharan uplands. If the eroded rocks were made of granite, the sand component derived from them would consist of the rock minerals quartz and feldspar, in roughly equal amounts. The feldspars would eventually weather to form clay. Quartz is resistant to chemical weathering and so tends to persist in the landscape when other rock minerals have been altered and either washed or blown away. If the parent rocks are sandstone, and there is a great deal of Nubian Sandstone of Cretaceous age within the Sahara, the sand derived from their erosion will also be quartz sand. Most of the Saharan sand consists of quartz, and in the case of Libyan dunes, the sand is almost pure quartz.[6]

Sand grains derived from the erosion of volcanic rocks like basalt will not be made of quartz. Volcanic sands are often black like their parent rocks. Many small volcanic plugs across the Sahara are simply called *Jebel Soda* in Arabic, or *Black Mountain*. The Arabic adjective *soda* attached to a hill or small mountain is a good guide to the volcanic origin of the bedrock.

Reworking River Sands into Sand Dunes

During the many drier climatic intervals, persistent desiccation ultimately deprived most of the Saharan rivers of water and they became choked in their own sediment. As the polar regions became progressively colder, and ice caps began to develop in high latitudes (see chapter 4, table 4.1), the temperature and pressure gradients between high and low latitudes steepened, the Hadley Circulation, which I discussed in chapter 4, became invigorated, and the Trade Winds became stronger

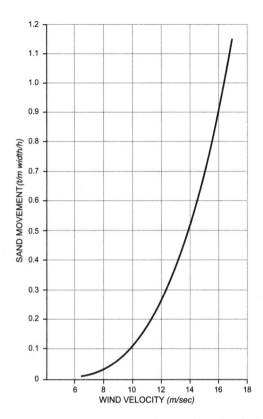

FIGURE 5.3. Wind velocity and sand movement. After M. Williams (2014). *Climate Change in Deserts: Past, Present and Future*. Cambridge University Press, Cambridge and New York, fig. 8.5. © Cambridge University Press. Reproduced with permission of the Licensor through PLSclear.

during these drier intervals. Marine sediment cores obtained off the west coast of the Sahara show a sudden increase in particle size and a substantial increase in the influx of Saharan wind-blown sand, all indicative of stronger Trade Winds at these times. Bagnold had demonstrated sixty years ago from his wind-tunnel experiments that the ability of wind to detach a grain of sand from the land surface and entrain it into the air increases exponentially with wind speed once a certain threshold wind velocity has been attained (fig. 5.3). Depending on particle size, the threshold value for sand mobilisation and transport lies between four and six metres per second.

Once the former alluvial sands had been mobilised by the desert winds, they were soon re-fashioned into a variety of dune forms.[7] The smallest of these are the *nebkhas*, which are simply small cones of sand formed around tussocks of grass or more elongated sand trails extending downwind of some small obstacle such as a boulder or a shrub. Somewhat larger than nebkhas are the tamarisk mounds that develop from sand accumulating around the tangled roots of tamarisk trees (plate 5.1). The resulting conical mounds are a useful guide for the desert traveller. Dig deep enough beneath the mounds and you will find water, although it may be slightly brackish.

On a much larger scale, and where there is an abundant supply of sand, are the great rounded whaleback dunes and the sinuous linear *seif* dunes (*seif* is Arabic for sword) that extend downwind for many kilometres more or less parallel to the direction of the dominant sand-moving winds. Such longitudinal dunes can be solitary, like the mighty Qoz Abu Dulu (fig. 5.1) which extends from north to south for more than 500 kilometres west of, and parallel to, the Nile; but more often they are concentrated within an extensive field of dunes known in the western Sahara as an *erg*. The highest dunes are not linear dunes but are pyramid-shaped and known as star dunes (plate 5.2), as in the Great Western Erg (fig. 5.1) and the Great Eastern Erg (fig. 5.1) of the northwest Sahara.[8] The dunes of the Great Sand Sea (fig. 5.1) of western Egypt and eastern Libya which Bagnold and his hardy companions eventually managed to drive through comprise both linear and more sinuous whaleback dunes. Individual dunes within these ergs can be over a hundred metres or more in height, particularly the star dunes. They usually have a relatively gentle slope on the windward side and a steeper slope or slip face on the leeward side. Figure 5.4 shows how dunes advance under the influence of a sand-moving wind. During our summer travels through the Libyan dunes of the Great Sand Sea in 1962 and 1963, I observed that the wind direction sometimes changed in late afternoon, so that what had started the day as a gentle windward slope

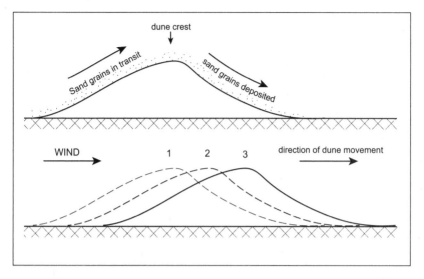

FIGURE 5.4. Dune advancing under the influence of a strong sand-moving wind.

now became an unstable and steep slip face down which loose sand grains went avalanching, which made for some interesting driving.

The general orientation of linear dunes across the Sahara as a whole is like a huge, flattened circular swirl, with dunes aligned according to the dominant anticyclonic wind regime (fig. 5.1). Monique Mainguet and her colleagues have shown how this wind pattern is responsible for moving sand from the northern and central Sahara to the southern and western areas of the desert.[9] On reaching an obstacle such as a mountain, the encroaching dunes diverge and move around the obstacle, like two outflanking limbs of an advancing army of sand (fig. 5.5). On occasion, I have seen sand dunes in Mauritania, Tunisia, Algeria, and Sudan (plate 5.3 and map 2) climbing up the sides of rocky cliffs and descending the opposite side. Wind, unlike flowing water, can move upwards with ease. However, strong winds are needed to generate these climbing dunes, as well as enough sand.

There is another well-known type of Saharan dune known as a *barchan* or *barkhan*. These dunes are crescent-shaped, with the crescent

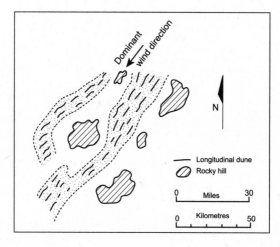

FIGURE 5.5. Dunes diverging around an obstacle.

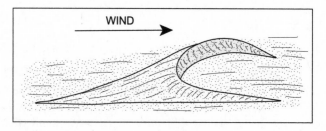

FIGURE 5.6. Barchan dune.

horns facing downwind (fig. 5.6). They are quite small, mostly only a few tens of metres in height, and are often remarkably mobile. They can occur in small groups, like a pod of desert dolphins, or alone in solitary splendour. I once slept atop an isolated barchan in the Libyan Desert in the summer of 1963, enjoying the feel of cooling sand after the sun had gone down. A strong wind came up that night and by dawn my sleeping bag was covered in sand, and to my delight I discovered that the slip face of the dune had already moved further downwind, as theory predicted.

Once they have become well-established, desert sand dunes will encroach across dry riverbeds and block their channels. Given enough time, they will obliterate all surface evidence of formerly flowing water.

The use of *spaceborne imaging radar*[10] in the early 1980s revealed a whole series of buried river channels in the eastern Sahara and proved an invaluable addition to the use of *Landsat* images when mapping desert landforms. Prehistoric stone tools dating back to the Lower Palaeolithic were found in trenches excavated across some of the more recent buried channels located using spaceborne imaging radar in the eastern Sahara. Some of the older channels are thought to date back to the Oligocene about 30 million years ago, which is quite possible, but they remain poorly dated.

The expansion of desert dunes across the Saharan lowlands had the effect of accentuating the consequences of aridity. During any ensuing wetter climatic intervals, the rivers faced an uphill battle to breach the dune barriers and re-establish an integrated drainage network. Some of the larger rivers proved successful, at least for a while, and allowed Mediterranean plants to migrate from the Atlas Mountains to the Saharan volcanic uplands of the Hoggar, Tibesti, and even as far south as Jebel Marra (map 1) in northwest Sudan. They also enabled tropical plants to migrate northwards into the central Sahara. Animals in due course followed the migrating plants and took up their abode in the heart of the Sahara. Human foragers followed these same routes to occupy the better-watered regions of the desert from about a million years ago onwards.

Renewed climatic aridity caused the cycle to repeat once more. Dunes again blocked the rivers; plants, animals, and prehistoric hunter-gatherers sought refuge in the uplands where water was reliable and food supplies adequate. The Saharan uplands have a long history of acting as ecological refuge areas, allowing plants and animals to weather the intervals of aridity until the return of dependable rain. Some of these plants and animals never again left their mountainous sanctuaries. The olive trees (*Olea laperrinei*) in the Aïr Mountains and in sheltered ravines around Jebel Marra are one example; so, too, are the relict populations of patas monkeys (*Cercopithecus patas*) and the baboons (*Papio anubis*) in the Aïr Mountains.[11] Their ancestors were originally from the tropical rain forests of West Africa when the rivers that once flowed south from the Aïr Mountains were flanked on either side by dense tropical riparian woodland. The dwarf crocodiles (*Crocodylus niloticus*) that lived in some of the permanent waterholes in the Tibesti massif until hunted to extinction in the 1950s were yet another relict population from previously wetter times.[12] In 1957 the British archaeologist Anthony J. Arkell made a journey from Kufra to Ennedi[13] across the Rebiana Sand Sea (map 3 and fig. 5.1). South of the shrunken lake at Wanyanga Kebir (map 3), in a valley called Wadi Zirmei, he found rock engravings showing the extinct elk-like *Megaceroides*, as well as elephant, rhinoceros, giraffe, antelope, ostrich, a possible lioness, hunting dog (*Lycaon pictus*), and cattle. In a gorge in the Ennedi massif (map 3) he commented that 'there is now in the gorge a series of five pools, with potholes in the upper reaches. There are fish in several of the pools, and crocodiles at least 6 ft (1.8 m) long.' Their presence was the last echo of a time when there was more water in this now arid landscape.

The presence of sand dunes was sometimes an advantage for prehistoric human settlement. Dunes will often store rainwater at depth, some of which may seep out along the flanks of the dune to form freshwater seeps and even sustain small ponds or wetlands between the dunes. Silt and clay sediments trapped behind a dune barrier can allow ponded water to accumulate because they are impermeable and so do not allow rainwater to infiltrate into the sand beneath. On the southern margin of the Hamada el Akdamin (map 3) sandstone plateau in the southern Libyan Desert, I dug a pit to a depth of 2.7 metres in the flat area between two dunes. The pit revealed stone scrapers and stone borers stratified within a loamy sand that had probably accumulated in a small seasonal lake or pond during the early Holocene at a time when the climate was less arid.[14] During these less arid phases, grass and trees will grow on the dunes but not on the rocky areas between the dunes and will help to stabilise the normally mobile dunes.

Do Desert Dunes Imply Aridity?

The southern limit of mobile desert dunes in the Sahara coincides remarkably closely with the 150 mm rainfall boundary or isohyet (fig. 5.7). Fixed dunes are found up to 500 kilometres, and in places nearly 800

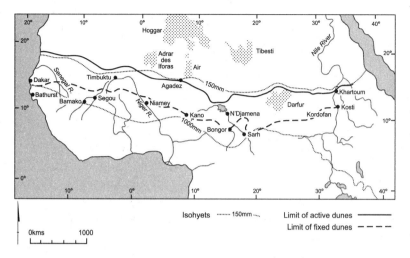

FIGURE 5.7. Map showing the limits of active and fixed dunes in and beyond the Sahara. The present-day limit of active dunes is bounded by the 150 mm isohyet. Fixed dunes extend up to 500 km south of the Sahara, locally into areas that now receive 1000 mm of mean annual rainfall. After M. Williams (2014). *Climate Change in Deserts: Past, Present and Future.* Cambridge University Press, Cambridge and New York, fig. 8.8. © Cambridge University Press. Reproduced with permission of the Licensor through PLSclear.

kilometres, south of the present-day limit of active dunes, as far south as some places that now receive 1000 millimetres of rain annually (fig. 5.7). Many of these now fixed dunes were mobile during the time of the last glacial maximum 20,000 years ago when a long interval of intertropical aridity was clearly evident in the marine sediment cores west of the Sahara that were studied by Michael Sarnthein and his team of marine geologists from Kiel in Germany.[15] The Saharan dunes were briefly stabilised by vegetation when wetter conditions resumed about 15,000 years ago, but became active once more between about 12,800 and 11,500 years ago during a cold and very dry climatic interval first recognised in Europe and known as the Younger Dryas.[16] *Dryas octopetala* is a wildflower that grows today in tundra and alpine regions in Eurasia.

For desert sand dunes to form, three things are necessary—a sufficient supply of sand, strong winds, and a sparse or non-existent plant cover. A dune does not necessarily denote aridity. If the winds are vigorous enough and the sand supply sufficiently copious, mobile dunes can develop even when there is significant rainfall. For example, the coastal dunes of the southern Negev Desert of Israel were actively advancing eastwards 20,000 years ago when the regional climate was far less arid than it is today, but the winds were stronger and there was an abundant supply of sand exposed along the coast.[17]

There are two distinct types of dune that are common along the semi-arid fringes of deserts, but that cannot form under conditions of extreme aridity. On the downwind margin of small, kidney-shaped lakes in the semi-arid zone of Morocco and Algeria there is a distinctive type of crescentic dune sometimes called a *clay dune* but known more generally as a *lunette* (fig. 5.8a). They form immediately next to the beaches of these seasonally fluctuating lakes and consist of sand-sized particles of clay detached from the dry and exposed portions of the lake floor during the dry season. These particles look like sand and behave like sand but are in fact sand-sized aggregates made of clay and in some cases of gypsum. If the climate became too arid, these lunette dunes would no longer be able to form.

The other type of dune that thrives in semi-arid rather than arid environments is the *source-bordering dune*. As the name implies, these dunes are located next to a source of sand which in most cases is river channel sand brought in by seasonal streams (fig. 5.8b). The three requirements to form a source-bordering dune are a regularly replenished supply of sand, river channels largely devoid of any obstructing vegetation along their banks, and, in the case of linear source-bordering dunes, strong unidirectional winds, at least seasonally. Once the rivers cease to flow, the dunes are deprived of their supply of sand and cease to grow.

Dunes, Desert Dust, and Fossil Soils

Winnowing by wind of the sandy alluvium brought down by rivers flowing from the Saharan uplands has a sorting effect. The finer silt- and clay-sized particles are blown away as desert dust, leaving a residue of sand and gravel. The sand particles are soon ferried away by desert

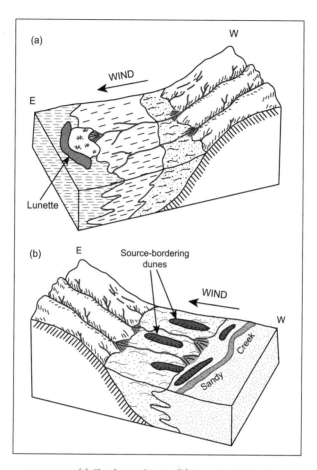

FIGURE 5.8. (a) Clay dune or lunette. (b) Source-bordering dune.
After M. Williams (2015). Interactions between fluvial and eolian
geomorphic systems and processes: Examples from the Sahara
and Australia. Catena Special Issue: *Landforms and geomorphic processes
in arid and semi-arid areas*, 134, 4–13, fig. 3. Reprinted with permission
from Elsevier.

winds, leaving a protective cover of fine and coarse gravel often called a desert pavement. Beneath these surface gravel layers which are usually only one stone in thickness there may be one or more buried soils. These soils can develop quite quickly, within a few score years. In such instances, the surface sands act as dust traps during times of rain or when the sand surface supports sporadic tussocks of grass. Once trapped beneath the sand surface, the dust particles, which often consist of calcareous clay, absorb any infiltrating dew or rainwater, and also trap grass seeds that can remain dormant and viable for many decades. Further plant growth traps more desert dust and the plants later decay to provide organic matter to the developing soil, which in turn allows the soil to trap and retain more moisture. Trenches dug beneath the desert pavements on the surface of flat areas of sand and below the surface of currently active sand dunes often reveal several generations of fossil soils sandwiched between layers of sterile desert sand. The soils denote slightly wetter intervals during which the dunes were relatively stable, and the sterile sands denote times when the desert sands were once more in motion.

Saharan Sandplains

There are vast tracts of the Sahara that do not consist of the familiar landscape of rolling dunes but instead consist of a very gently undulating surface of wind-blown sand often interspersed at shallow depth with buried fossil soils. These vast plains are often called sand sheets or sandplains because of their vast extent and very gently undulating topography. The great Selima Sand Sea[18] (fig. 5.1) in southwest Egypt and northwest Sudan covers an area of about 120,000 square kilometres and is today one of the driest parts of the eastern Sahara. It was not always quite so arid.

American geo-archaeologist C. Vance Haynes spent many years carefully gathering evidence from fossil soils, prehistoric stone tool assemblages, and fossil pollen grains recovered from the deposits that accumulated on the floor of former lakes within this region to show that the rainfall belts that now lie well to the south in central Sudan once reached as far north as Selima Oasis[19] (map 3). This interpretation, if correct, would mean a northward shift of the rainfall belts by about 500 kilometres during a wet interval that occurred between about 8500 and 6000 years ago. He called the now desolate sand sea that includes both the gently undulating *Selima Sand Sheet* to the east and the rolling dunes of the *Rebiana Sand Sea* in the Libyan Desert to the west the *Darb el Arba'in Desert*. The name refers to the notorious forty days' slave route (or *Darb el Arba'in*) running from northwest Sudan to the Nile in Egypt. The route was not entirely devoid of water. There was fresh water close to the surface at Bir Sahara and Bir Tarfawi (map 3) and available from shallow wells. Selima Oasis also had permanent water. Nevertheless, scattered along the route one often comes across the almost complete skeletons of camels bleached white by the sun—mute witnesses to the incredible hardships of the infamous journey.

Just how these vast, nearly horizontal sandplains originate is still shrouded in mystery. One possibility favoured by geologist Mike Talbot is that they represent the final stage of prolonged erosion by both wind and water of a former field of desert dunes, with the ultimate surface lying close to the underlying water table at the time of its final phase of wind deflation.[20] Another possibility is that they represent the final stages of infilling of former topographic depressions, with wind and water both contributing to the infilling process. Whatever their origin, they are today among the least hospitable parts of the Sahara.

Wind or Water?

It is all too easy to underestimate the role of water in fashioning the Saharan landscapes. The constant soughing of the wind, the seemingly unstoppable advance of active dunes, the burial of former river channels and lakebeds beneath a deep mantle of wind-blown sand, all testify to the power and presence of wind in the desert. Occasional heavy rain-storms generally do little to dispel this image. The exceptional rains at the end of 1969 and the start of 1970 certainly brought much needed relief to the oases of Biskra and Touggourt (map 1) in Saharan Algeria by helping replenish the shallow aquifers on which the people depended for water and, except in the coastal mountains, did little damage. Among the inland sand dunes of Mauritania, the rain clouds of January 2004 conjured up a magnificent sunset (plate 5.4) and revived the grass cover in sheltered spots between dunes, but did little more.

However, in the heart of the eastern Sahara there is clear and unambiguous evidence of the power of running water to erode deep valleys. Consider for example the isolated mountain called Jebel Arkenu (map 1 and fig. 5.9) located in the hyper-arid southeast Libyan Desert. Arkenu is a small ring-complex, which means that when it first formed, volcanic rocks penetrated through the overlying sandstone sedimentary cover in a series of concentric rings of volcanic rock known as ring-dykes. This took place between about 50 and 30 million years ago, during the Eocene. Some of these rocks proved resistant to erosion and form steep-sided, boulder-mantled circular or elliptical ridges. Between them, the least resistant of the rocks form narrow valleys enclosed by the resistant ridges.

Cut almost all the way through the core of the mountain is a valley (see fig. 5.9) which I call the Main Wadi.[21] It is entrenched nearly a thousand metres into the mountain, which rises to an elevation of 1435 metres above sea level. This ravine was eroded entirely by running water. Walking upstream from the mouth along the dry channel bed of the Main Wadi you are flanked on either side by two alluvial terraces (see fig. 5.10). The upper terrace is between five and seven metres above the wadi floor and consists of rounded boulders often larger than 30 centimetres in diameter within a matrix of coarse sand and fine gravel. On the surface of this gravel terrace there are a few waste flakes and fragments of pottery, probably of Neolithic age. The lower terrace is only about two metres in height and is made up of boulders 15–30 centimetres in diameter. The stream that deposited both sets of gravel was obviously capable of significant erosion and must have been flowing rapidly to transport such large boulders. Further upstream you find yourself walking up narrow tributary ravines with alternating steep and gentle reaches. The gentle reaches become pools during rare rainstorms and are up to five metres deep and twelve metres long. In one instance, where the wadi cut across a very hard outcrop of rock, the 'pools' were deep, closely spaced, and bounded by overhanging ledges.

FIGURE 5.9. Google Earth black and white satellite image of Jebel Arkenu, Libya.

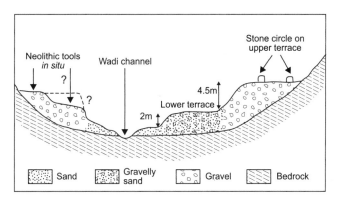

FIGURE 5.10. River terraces in the main wadi at Jebel Arkenu. After M. Williams (2019). *The Nile Basin: Quaternary Geology, Geomorphology and Prehistoric Environments*. Cambridge University Press, Cambridge and New York, fig. 16.2. © Martin Williams. Reproduced with permission of the Licensor through PLSclear.

An alpine shepherd accustomed to mountain torrents would have recognised the wadi for what it really is—a mountain torrent during times of vastly increased rainfall and torrential stream flow. Jebel Arkenu was fashioned almost entirely by the erosive power of running water. A solitary boulder fifteen centimetres in diameter located 600 metres from the mouth of the Main Wadi is a further reminder of the power of running water, even on slopes as gentle as two degrees. Nor is Jebel Arkenu an isolated example. Its bigger and better-known cousin, Jebel 'Uweinat ('mountain of small springs') (map 1), with its four permanent springs and over a thousand rock paintings and engravings, is also a ring-complex and also originated during the Eocene. It is twenty-five kilometres in diameter and rises to an elevation of 1,934 metres, flanked by the surrounding sand and gravel plain which lies about 450 metres above sea level. Like Arkenu, 'Uweinat is deeply dissected.

What we are looking at here is a rate of vertical downcutting amounting to about forty metres per million years, which is far higher than present-day rates of denudation on hard rocks in arid regions. The clue to this rapid erosion comes from the highly weathered state displayed by many of the rocks at Arkenu and 'Uweinat, as well as in other Saharan ring-complex mountains such as Adrar Bous to the east of the Aïr Mountains. Prolonged and deep chemical weathering during Eocene times, when what is now the Sahara was then covered in tropical rain forest, produced soft and friable weathering mantles that were easily eroded when conditions became drier. It is safe to conclude that although there is a constant tug-of-war between wind and water in the present-day Sahara, the major elements of the Saharan landscape were carved from the rocks by running water long before the first wind-blown sand made an appearance. It was the now mostly defunct Saharan rivers that supplied the parent alluvial sands that were then fashioned into dunes as the desert winds grew stronger and the overall climate became more arid. In chapter 6 we consider what happens to the desert dust blown from the Sahara.

CHAPTER 6

A Handful of Dust

I will show you fear in a handful of dust.

T.S. ELIOT, *THE WASTE LAND*

The Amazon Rain Forest Is Fertilised
by Saharan Desert Dust

It is quite amazing to think that wind-blown dust from the Sahara can have a beneficial influence upon the rain forests of the Amazon basin today, yet it is so. The Saharan desert dust that reaches as far as the Amazon basin (fig. 6.1) plays an important and possibly even a critical role in sustaining the health of the Amazon rain forests. The dust particles provide nuclei for the formation of ice crystals in clouds above the rain forest and so help to enhance or maintain precipitation over the Amazon forests.[1] Equally important, trace elements within the dust such as nitrates, phosphorous, and potassium are a major source of plant nutrients.[2]

This fertilisation process has probably been in operation at least intermittently for up to seven million years. We saw in chapter 4 that the Sahara began to show signs of aridity by about seven million years ago. Wind-blown sands are evident in the Chad Basin at about that time and evaporite deposits indicative of aridity began to form in the far northwest of Africa from about that time onwards. The sparse pollen record available from Saharan Miocene sediments also points to incipient

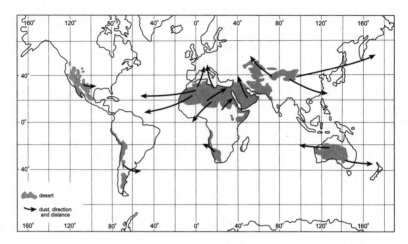

FIGURE 6.1. Map of major Saharan dust sources and directions of transport. After M. Williams (2014). *Climate Change in Deserts: Past, Present and Future.* Cambridge University Press, Cambridge and New York, fig. 9.1. © Cambridge University Press. Reproduced with permission of the Licensor through PLSclear.

aridity,[3] as do Atlantic marine sediment cores collected close to the western margin of North Africa.[4]

It is something of a paradox that it is during the most arid intervals, when the Sahara is largely devoid of vegetation and dust storms are most intense, that the Amazon forests benefit most from fertilisation provided by the influx of Saharan dust. It is also during these cold, dry climatic intervals that the rain forests in both the Congo/Zaïre basin and the Amazon basin were much reduced in area[5] relative to their much greater extent during the warm, wet climatic intervals. What, then, is desert dust?

What Is Desert Dust?

Desert dust[6] consists of very fine mineral and organic particles and can form in many different ways. Perhaps the most common process occurs when rivers flowing from the Saharan uplands deposit a mixed load of gravel, sand, and silt in depressions next to the uplands. As we saw in chapter 5, the winnowing action of strong winds fashions the alluvial sands into dunes and sandplains while the silt particles are blown much further afield in the form of desert dust.

Breakdown of rock through salt weathering is one way to form fine mineral particles from weathered rock and was observed in action on the Egyptian pyramids by the Greek historian Herodotus (ca. 485–425 BC) some 2500 years ago.[7] In Book Two of *The Histories*, Herodotus states that he has 'noticed how salt exudes from the soil to such an extent that it affects even the pyramids.' As water evaporates upon reaching the desert surface, small amounts of dissolved salt form into crystals which then act as wedges and dislodge small fragments of rock. This process of salt weathering has been replicated in many laboratory experiments.

Another way to produce fine particles is by glacial action, when rocks grind against each other during glacier flow. The resulting 'glacial flour' consists of silt-sized particles that are relatively easy for strong winds to detach and transport once they have dried out. Such dust storms are common in the glacial outwash fans of Alaska and Iceland today. Both the Atlas Mountains north of the Sahara and the Hoggar Mountains in the central Sahara were glaciated as recently as 20,000 years ago, so that some desert dust could result from glacial action.

A further mechanism which, rather surprisingly, has only recently been confirmed, following observations in the Sinai Desert of northeast Africa and the Negev Desert of southern Israel, is the effect of mechanical abrasion of sand grains carried in suspension by strong turbulent winds.[8] As the larger sand grains crash into each other, smaller fragments of quartz are broken off and carried aloft in suspension as desert dust.

The fine sediments exposed at the surface when lakes and wetlands dry out are another important dust source. Strong winds charged with abrasive sand grains scour out the fine grains from the surface of dry lakebeds and often form deep grooves parallel to the wind direction, some of which may be several metres deep and many hundreds of metres long. Isolated rock outcrops in the Sahara are also often undercut at the base by wind abrasion.

Soils developed during wetter climatic intervals are another potential source of desert dust. As the friable and organic-rich surface layers of soil dry out, they become susceptible to erosion by both wind and water. The organic matter derived from former soils, wetlands, and lakes provides the nutrients that help to sustain the rain forests of the Amazon basin today.

Saharan Dust Blows as Far as Sweden

Dust storms are a fact of life for the people who live in and around the Sahara and have been so for as long as people lived in or near the desert. The British climatologist W.G. Kendrew[9] considered that 'dust, not rain, is the great discomfort of life in arid lands. Except on still nights the air is full of fine particles which percolate through the finest chinks into houses and even closed boxes. Dust lies thick on every shelf, covers furniture, settles on food, and is inhaled in the air we breathe'. In his powerful novel *The Grapes of Wrath* (1939), John Steinbeck gives a similarly vivid account of the discomfort endured by Oklahoma farmers and their families during the time of the great drought in the 1930s, which gave rise to the term Dust Bowl.

On rare occasions, desert dust blown from the Sahara reaches as far as northwest Europe. In July 1968, Saharan dust was recorded across Britain.[10] In Sweden people awoke in alarm one winter morning on March 10, 1991 to find the snow around their houses covered in a strange yellow substance[11] which later proved to be Saharan dust, to their immense relief. On another occasion thirty years ago, at Cork airport in southern Ireland, heavy rain brought down Saharan desert dust from high in the atmosphere together with a well-travelled live Saharan locust. Since then, the Irish press has reported several more instances when invasions of live Saharan locusts have made conditions precarious at both Cork and Dublin airports.

These dust outbreaks into Europe are quite different from the plumes of dust that blow from the Chad Basin out across the Atlantic when the Harmattan wind is blowing during the dry season. In fact, there are at least four major dust trajectories operating today over North Africa.

PLATE P.1. Shoreline of a former lake at Adrar Bous in the heart of the Sahara. The irregular brown material on the surface of the sandy beach-ridge consists of fossilised reeds. Shells of freshwater snails are scattered through the beach sands.

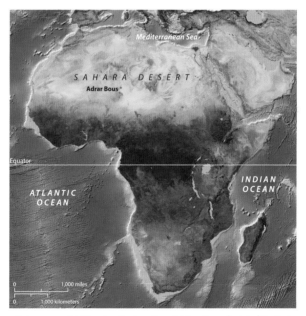

PLATE P.2. Colour map of Africa showing location of Adrar Bous (red dot) in the heart of the Sahara. Green denotes vegetation cover; yellow denotes lack of vegetation. Simplified from *The Times Atlas of the World*. (1980). Times Books, London.

PLATE P.3. Our Tuareg guide, Zewi bin Weni, riding his camel on the way to Adrar Bous in the south-central Sahara.

PLATE P.4. Desert dune bordering a former river channel immediately east of the Aïr Mountains.

PLATE P.5. Ancient river flood plains now exposed as alluvial terraces in the flanks of a desert dune east of the Aïr Mountains.

PLATE A.1. The sand dunes of the Ténéré Desert, central Sahara. Photo by J. Desmond Clark.

PLATE A.2. Rugged mountain range in the Aïr Massif, central Sahara.

PLATE A.3. Gravel-covered surface in the central Sahara.

PLATE 1.1. Wind erosion has caused undercutting of this small sandstone knoll in the central Sahara.

PLATE 1.2. Volcanic plug near Tamanrasset in the Hoggar Mountains, Saharan Algeria.

PLATE 1.3. The Hermitage built by Père Charles de Foucauld in 1910 at Assekrem near Tamanrasset in the Hoggar Mountains, Saharan Algeria.

PLATE 1.4. El Berbera Oasis, Mauritania desert, western Sahara.

PLATE 3.1. Fossil jaw and vertebrae of Nile perch in the desert of northern Sudan. Geological hammer is 28 cm long.

PLATE 3.2. Floor of former lake with desiccation cracks filled with sand and dunes encroaching on the dry lakebed, desert of northern Sudan.

PLATE 3.3. Adrar Bous, an isolated mountain in the south-central Sahara.

PLATE 3.4. Skeleton of a 5000-year-old short-horned Neolithic domestic cow (*Bos brachyceros*) at Adrar Bous.

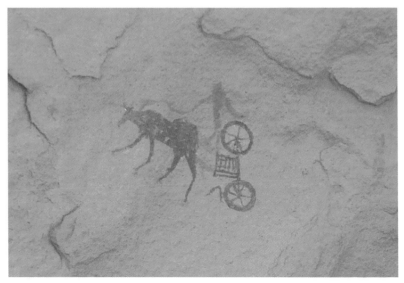

PLATE 3.5. Painting of two-horse chariot, southwest Libyan Desert. Photo by, and courtesy of, Claudio Vita-Finzi.

PLATE 3.6. Rock engraving of a giraffe, southwest Libyan Desert. Photo by, and courtesy of, Claudio Vita-Finzi.

PLATE 3.7. Rock engraving of an elephant, southwest Libyan Desert. Photo by, and courtesy of, Claudio Vita-Finzi.

PLATE 3.8. Rock engraving of an antelope, southwest Libyan Desert. Photo by, and courtesy of, Claudio Vita-Finzi.

PLATE 3.9. Neolithic rock painting of women riding oxen, Iheren, Tassili, Algeria. Copy by P. Colombel, Mission H. Lhote 1970. Archive Musée national d'Histoire naturelle and Musée de l'Homme, Paris. Reproduced from a photograph with permission of Dr Rudolph Kuper, University of Cologne.

PLATE 3.10. Rock engraving showing two possibly mythical creatures sparring, southwest Libyan Desert. Photo by, and courtesy of, Claudio Vita-Finzi.

PLATE 5.1. Dune formed around former tamarisk roots.

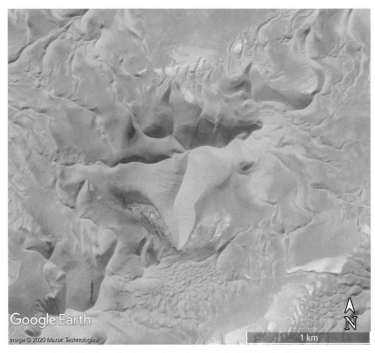

PLATE 5.2. Google Earth satellite image of star dunes, Erg Iguidi, Algerian Sahara.

PLATE 5.3. Sandstone hill with climbing dune, northern Sudan.

PLATE 5.4. Desert sunset before the storm, Mauritania.

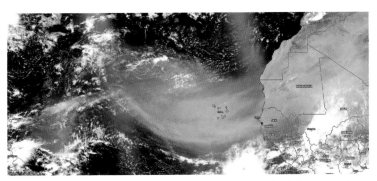

PLATE 6.1. NASA image of 18 June 2020 dust plume blowing in a curved path from the northwest Sahara across the Atlantic.

PLATE 6.2. Wind-eroded remnants of former lakebeds, northern Sudan.

PLATE 6.3. Modern desert stream channel in the Matmata Hills of Tunisia cut into fine-grained sediments derived from reworked wind-blown dust. The cliff is about 10 metres high (see person at foot of cliff for scale).

PLATE 7.1. Early Stone Age biface or hand-axe from the central Sahara.

PLATE 7.2. Broken grindstone in the desert of northern Sudan.

PLATE 7.3. Google Earth satellite image showing present-day Lake Chad flanked by fixed dunes which migrated across the floor of the former much enlarged Lake Chad, which reached its Holocene maximum about 9000 years ago.

PLATE 8.1. The bare, cracked surface of a clay soil during drought.

PLATE 9.1. Remains of former Roman dam built across an ephemeral stream channel in the Tunisian desert in order to trap silt and soil water to enable olive trees and date palms to be grown.

PLATE 9.2. Modern porous stone dam built across an ephemeral stream channel in the Tunisian desert allowing silt to be trapped, moisture to be stored in the silt, and date palms and olive trees to be grown.

PLATE 10.1. Afar women filling goatskin bags from a shallow well dug in the dry floor of Lake Lyadu, Afar Desert, Ethiopia.

One, which caught Dr Matthew Dobson's attention in 1781,[12] is the Harmattan dust-bearing wind which blows westwards along the southern margin of the Sahara, across the Atlantic to Barbados and then across the equator as far as the Amazon basin in South America.

Another dust pathway (plate 6.1), which caught the attention of Charles Darwin[13] in 1832, flows in a curved path west across the northwest Sahara to the Atlantic and finally to North America.[14] A third dust route is from the central and northern Sahara eastwards across Egypt and the Mediterranean to reach Israel and Arabia. A fourth major dust path is northwards across the Mediterranean to the Alps and southern France, where it contributes to the vineyard soils of that famous wine-producing region. Dust plumes from the northern Sahara sometimes reach as far north as England, Ireland, Sweden, Finland, and even Iceland.

On the island of Crete, which was previously joined to the Greek mainland, there are footprints of a bipedal hominin-like creature preserved in silty clay deposits that originated as Saharan desert dust some 5.7 million years ago.[15] On mainland Greece, at the site of Pyrgos, a fossil jaw of a possible hominin, *Graecopithecus freybergi*, has been recovered from another deposit of ancient Saharan desert dust with an age of 7.2 million years.[16] It thus seems very likely that Saharan desert dust has been contributing to the soils of what are now the olive groves and vineyards of southern Europe for the past seven million years.

The 'Khamsin' or 'Fifty-Day' Wind and Its Impact

The winds responsible for detaching desert dust particles and transporting them across the Sahara have given rise to a number of picturesque local names, including *ghibli*, *sirocco*, *haboob*, *harmattan*, and *khamsin* (pronounced 'hamseen'). The khamsin wind is a hot, dry wind that is common between February and June in Egypt and Libya. It can reach wind speeds well above one hundred kilometres per hour and can transport substantial quantities of sand and dust. The name comes from the Arabic word for fifty—an indication of how persistent these winds can be. Napoleon found the khamsin sandstorms extremely trying during his campaign in Egypt.

These winds are important for their role in picking up and transporting desert dust. Certain conditions are necessary for dust mobilisation, including aridity, a sparse plant cover, and a broad expanse of detachable soils or sediments. The active agents of particle entrainment are strong, turbulent, and gusty winds. Without such winds there is no dust detachment. Once suspended in the atmosphere, where convectional air currents can carry them well aloft, the dust particles can be carried by strong unidirectional winds far beyond their original source area.

Abundant remnants of former lakes are found in the Western Desert of Egypt to the south of Dakhla Oasis and Kharga Oasis (map 3). The deposits consist of calcareous silts that had gradually accumulated on the bed of these ancient lakes during times of wetter climate. The silts contain fossil shells of aquatic snail shells and the calcareous shells of ostracods. Ostracods are very small crustaceans. Today these former lake floor sediments no longer occupy low points in the landscape but stand proud as a result of wind erosion of the land that once surrounded the lake (plate 6.2). This relief inversion is very common in areas where the winds are strong and highly abrasive.

A similar, equally spectacular inversion of relief is evident all the way south to the two prehistoric sites known as Bir Sahara and Bir Tarfawi (map 3) which have yielded a rich trove of Middle Stone Age artefacts and fossil bones to the trowels and brushes of the Combined Prehistoric Expedition directed by Fred Wendorf and Romuald Schild.[17]

Former lakebeds are not the only landform prone to this type of relief inversion under the influence of powerful winds like the khamsin. Ancient and long defunct river channels in the Western Desert of Egypt now stand aloft in the landscape as well, and have their counterparts on Mars, which has a similar set of what appear to be former river channels that are now elevated above the surrounding Martian terrain.[18]

Herodotus and the Libyan War on Saharan Winds

In this section we discuss some of the more tragic impacts of the khamsin winds in Libya and Egypt. Sandstorms are a bane to any desert traveller and can be lethal. They are to be endured rather than fought against. In Book Four of *The Histories*, Herodotus (ca. 485–425 BC) tells the story (perhaps quite apocryphal) of a group of people from the small town of Sirte in northern Libya who were enraged by the south wind (most likely the khamsin) because it had dried out the water in their cisterns.[19] In a quite irrational fury, the people of Sirte declared war on the wind and marched into the desert, where 'the wind blew and buried them in sand'.

Later in Book Four, Herodotus describes 'salt-hills and springs' in Libya and noted that 'the houses are all built of salt-blocks—an indication that there is no rain in this part of Libya, for if there were, salt walls would collapse. The salt which is mined there is of two colours, white and purple. South of the sand-belt, in the interior, lies a waterless desert, without rain or trees or animal life, or a drop of moisture of any kind'.[20] When I visited Kufra Oasis (map 3) in southern Libya in 1962 and 1963, the main road was made of rock salt, a clear sign that rain was very rare.

Another example cited by Herodotus involved the Persian army of 50,000 soldiers led by King Cambyses II, the son of Cyrus the Great. They were on their way to quell a rebellion in Siwa Oasis (map 3) in the Western Desert of Egypt in about 524 BC when they were caught in a sandstorm and were never seen again. In the account reported to Herodotus, 'a southerly wind [again, very probably the khamsin wind] of extreme violence drove the sand over them [the Persian soldiers] in heaps as they were taking their mid-day meal, so that they disappeared for ever'.[21] In light of the engrained memory of these ancient disasters, it is small wonder that in his great poem *The Waste Land*, T.S. Eliot wrote, 'I will show you fear in a handful of dust'.

However, as we saw in chapter 3, life in the desert has not always been as grim as it was during the time when the army of Cambyses II was engulfed in sand. Only a few thousand years before Herodotus visited Egypt, numerous small bands of pastoralists grazed their herds of cattle across what is now the arid Sahara. They left behind a legacy of rock paintings depicting Neolithic cattle, sheep, and goats. Earlier still, as we saw in chapter 3, the prehistoric hunters of the Sahara carved rock engravings showing antelopes, giraffes, elephants, and other large animals that live today in the savanna lands of East Africa. In those times the

Sahara was a well-watered land dotted with small lakes and perennial rivers and well able to support a savanna vegetation of scattered woodlands and grasslands.

The Great Dust Storms in the Nile Valley

Not all dust-bearing winds blow with the persistence and regularity of the khamsin or the Harmattan winds. In the Nile valley of southern Egypt and northern Sudan there are sporadic seasonal dust storms known as *haboobs* which turn day into night and make life hard for humans and their animals.[22] Such dust storms are highly turbulent and can be at least 1500 metres high and 25–100 kilometres wide. They can advance at speeds of 50–100 kilometres per hour. During the dry season, the wide flood plains of the Nile and surrounding sand dunes provide these gusty winds with an abundant supply of sand and finer particles. In Arabic, the word haboob has two somewhat different meanings. One denotes blasting, as in sand blasting; the other denotes drifting, as in sand or silt moving across the landscape. Both are apt descriptions of the haboob dust storms.

Climatologists J.F. Griffiths and K.H. Soliman[23] define a dust storm as 'a rapidly moving mass of air containing large amounts of dry, opaque particles which reduce visibility to less than 1000 m', and they have identified several types of haboob which blow at different times of the year and are controlled by regional weather conditions. Early geologists working in Sudan were so impressed by the amount of silt mobilised by northerly winds and later deposited on the clay plains of central Sudan that they interpreted these plains as being formed by wind-blown dust, much like the great loess plains of North America, northern Europe, and western China. They were in fact incorrect: these plains consist of fine alluvial sediments laid down over many thousands of years by the great Blue and White Nile rivers and their tributaries, but these soils do contain a modest component of silt derived from wind-blown dust. In contrast to soils derived solely from desert dust, the alluvial soils between the Blue and White Nile rivers contain shells of many species of aquatic snails as well as shells of freshwater oysters and mussels.[24]

Charles Darwin and His Dust Samples

Chinese scholars have studied dust storms and their associated deposits for well over a thousand years in China,[25] but European scholars only began to take a serious interest in dust storms in the last couple of centuries. Mariners crossing the Atlantic were well aware of the great plumes of dust coming from northwest Africa that were blown far out to sea during the northern winter season. In 1781, Dr Matthew Dobson[26] described the role of the Harmattan wind in transporting dust across West Africa and out to the Atlantic. His general observations have been confirmed ever since, including by Charles Darwin.

When HMS *Beagle* was anchored at Porto Praya in the Cape Verde archipelago off the west coast of the Sahara in January 1832, Darwin collected a sample of wind-blown dust derived from the Sahara by using a gauze filter placed at the ship's masthead. Darwin had also asked his geologist friend Charles Lyell, who was in a vessel several hundred kilometres further north, to collect additional samples for him. He sent his own sample together with the four dust samples collected by Lyell to the distinguished German naturalist Professor Christian Gottfried Ehrenberg in Berlin.

Ehrenberg had a long-standing interest in Saharan desert dust carried across the Atlantic and had written scholarly accounts of what he called *Blutregen* or 'blood rain' from the colour produced by iron oxide–coated red dust.[27] Ehrenberg managed to identify sixty-seven different species of diatoms in the five samples he received, two of them marine and the rest freshwater.

Diatoms are microscopic algae that live in both fresh- and salt-water and in damp soils. The outer portion of each individual diatom consists of a protective cover made up of silica, known as a frustule. Each frustule has a distinctive pattern of ornamentation which is often very beautiful. These frustules are quite resistant to weathering and can persist within layers of diatomite for many thousands of years. Diatoms were frequently very abundant in the lakes that dotted the Sahara between about 15,000 and 5000 years ago. Once the lakes had dried out, the friable deposits of diatomite or diatomaceous earth that had accumulated on the former lake floor were exposed to wind erosion and formed a ready source of potential desert dust.

'The Dustiest Place on Earth'

Both Dr Matthew Dobson in 1781 and Charles Darwin in 1832 were well aware that winds blowing westwards from the Sahara were responsible for ferrying dust out across the Atlantic. As we saw earlier, the Harmattan wind is a seasonal wind that blows from about November to March from the Chad Basin across northern Nigeria and out into the Atlantic. Many of the diatoms observed nearly two centuries ago by Darwin and later studied in detail by Ehrenberg most likely came from now dry lake sediments in the Sahara, including from around the Bodélé Depression in the Chad Basin (map 1). This locality has been aptly described as 'the dustiest place on earth' and has been the focus of intense study to determine why wind erosion is so severe in this locality.[28] The dry north-easterly winds that blow through this region generate about a hundred major dust plumes each year. The velocity of these winds increases when they are funnelled through a gap between the Tibesti and Ennedi (map 1) mountains and act as a low-level easterly jet stream[29] before they reach the Bodélé Depression. Here the turbulent winds cause the friable diatomites to disintegrate into small fragments that are easy to detach and carry away, especially because diatomite is far lighter than quartz. The dunes in the vicinity of the former lakebeds are a further source of dust, this time in the form of very fine particles of quartz. Without the seasonal presence of the strong easterly winds, there would be no Atlantic dust plumes, and Darwin would not have been able to trap his dust sample. During drier intervals in the past, the amount of dust exported from this region would have been even greater than it is today. Extreme aridity during the Last Glacial Maximum some 20,000 years ago would have meant that vast areas in the Chad Basin were exposed to wind erosion, a situation that persisted until about 15,000 years ago, when Lake Chad refilled and a very deep but fluctuating lake persisted until about 5000 years ago, when precipitation decreased and rivers brought less water to the lake.

Dust Recycling

We saw that desert dust can be derived from wind-blown river sediments. However, we tend to forget that river sediments can in their turn be derived from desert dust. In the arid limestone country of the Matmata Hills of central Tunisia (map 1), the modern ephemeral streams flow only rarely. When they do flow, they carry a coarse load of limestone cobbles, which is entirely consistent with the occurrence of occasional flash floods in this desert environment. However, the vertical banks flanking these ephemeral stream channels consist of very fine sediments (plate 6.3) which were laid down under very different conditions. They consist of wind-blown desert dust that was deposited across the landscape and washed down into the valley bottom before the onset of the present arid climate. Similar very fine-grained sediments occupy valleys in other desert areas, including the Sinai Desert, and also consist of wind-blown dust reworked by running water.

In the semi-arid high plains at the foot of the Aurès Mountains (map 1) in northern Algeria, there is a different type of desert dust. It consists of very fine, highly cohesive gypsum[30] blown from the surrounding gypsum dunes and former lake sediments rich in gypsum crystals. Some of the wind-blown gypsum dust has been partly reworked by runoff from the adjacent mountains. The resulting plains are now covered in saltbush. During times of war, the local people dug underground shelters that were easy to camouflage and remained stable thanks to the highly cohesive nature of the gypsum soil.

Saharan Dust in Atlantic Marine Sediments: A Tale of Wetter and Drier Times

We know from direct observations that years of severe drought along the southern Sahara and the Sahel are followed by an increase in the amount of desert dust blown from the southern Sahara out into the Atlantic.[31] It is therefore quite logical to expect that during drier times in the past there also would have been more desert dust blown out to sea. This is entirely to be expected and is true of deserts across the globe. Glacial intervals in the past have almost always been much drier in the tropical desert regions of the world. Deep-sea sediment cores show that the dust flux was three to five times greater during glacial times than during interglacial times. One reason for this concerns the difference in temperature and atmospheric pressure between the equator and poles. As the poles became progressively colder during the course of a glacial cycle and sea-surface temperatures in high latitudes became colder, the temperature and pressure difference between equator and poles increased. As a result, the Trade Winds[32] blew more strongly and were better able to mobilise and transport sand particles from the Sahara and other great tropical deserts, fashioning them into desert dunes. The finer particles were blown out to sea. We can therefore use changes in the amount of dust recorded in marine sediment cores off the coast of the Sahara as an indication of past climatic conditions over the desert. Times that were wetter, such as between about 15,000 and 5000 years ago, had a reduced dust flux, as Peter deMenocal and his colleagues have shown in a pioneering study,[33] and times that were windier and drier, such as about 25,000–15,000 years ago, had a much-increased dust flux. Because many factors are responsible for mobilising and transporting desert dust, it is not wise to read too much into the dust signal. It is also not wise to generalise about the climate over the whole of North Africa on the basis of only a single marine sediment core. If we have learnt anything during the last few decades about past climatic fluctuations in and around the Sahara, it is that the response of landforms and plant ecosystems to past climatic changes in this region has been complex, often counterintuitive, and not always taking place at the same time.

The next chapter looks at how prehistoric humans coped with living in the Sahara.

CHAPTER 7

Wood-Smoke at Twilight

Who has smelt woodsmoke at twilight? . . . Who is quick to read the
noises of the night?

RUDYARD KIPLING, *THE FEET OF THE YOUNG MEN*

Ancestral Memories

A room in the British Museum in London is home to clay tablets on
which are engraved in cuneiform script a vivid description of a huge
flood. The story revealed by the tablets, popularly known as the *Epic of
Gilgamesh,*[1] was inscribed between about 4000 and 3500 years ago in
Uruk in Mesopotamia near the banks of a former channel of the Euphrates River. The Old Testament account in Genesis of Noah's flood is
even more widely known and had a huge impact on nineteenth-century
geological thinking. The so-called diluvial or 'flood' deposits in England,
Ireland, Wales, and Scotland, which we now accept as sediments laid
down during the retreat of the ice caps that covered much of the British
Isles 20,000 years ago, were long regarded as evidence of a catastrophic
flood.[2] So entrenched was this view that William Dean Buckland[3]
(1784–1856), the eccentric first Reader in Geology at the University of
Oxford, failed utterly in his efforts to persuade a disbelieving audience
of eminent geologists that diluvial sediment was glacial, not fluvial.
Buckland had earlier coined the term *diluvium* or *flood* deposit to dis-
tinguish it from everyday alluvium. It was at the end of a lecture he gave

in November 1840[4] to the eminent members of the Geological Society at Burlington House in London that he was reputedly so enraged by the hostile reception to his new-fangled ideas (he was but a recent convert to glaciation) that he inveighed against his audience with characteristic vigour, concluding that if there was any man present [there were no women in the audience] who did not believe the evidence presented, 'may he be afflicted with eternal itch without benefit of scratching!'

Ancestral memories of land being flooded are common in many coastal regions of the world.[5] For example, the clan elders of the tropical coastal settlement of Milingimbi in the Northern Territory of Australia often fish in the Arafura Sea, and one day they remarked that in earlier times they used to live 'down there', pointing to the sandy seabed in these shallow waters. In fact, the sea reached its present level in these parts between 6000 and 7000 years ago, so they had kept alive these ancient memories in song, dance, and oral tradition for a very long time.[6]

At the height of the last glacial maximum 20,000 years ago, so much water was locked up in the form of ice that global sea level was about 125 metres lower than it is today.[7] Once the ice caps began to melt, which they did, rapidly at first, then more slowly, sea level rose in tandem, rapidly for the first few thousand years, then more slowly. In those regions of the world where the continental shelves are wide and the seas above them quite shallow, the horizontal loss of land during the initial stages of sea level rise could amount to about three feet or a metre a week, with submergence of 250 miles/400 kilometres of once prime coastal land over a period of about 10,000 years.[8] Small wonder that loss of land from flooding was etched in ancestral memory.

Memories of past migrations are also part of our ancestral memories. The great exodus of the Children of Israel from Egypt is a prime example. Whether forced out of their ancestral homes by war, famine, or pestilence, faint memories of long walks to more promising lands have been with us for many thousands of years. In fact, our remote ancestors have been on the move out of Africa again and again for more than a million years and have on occasion returned, bringing back new ideas and new stone tool kits with them. The move out of Africa would have involved some of our ancestors crossing the Sahara and Sinai deserts, leading us to ask when such crossings might have occurred, and under what sort of conditions. Before we do, we need to look at who were these remote ancestors of ours, and where did they originate.

The Dawn of Humanity

In his book *The Descent of Man*, Charles Darwin[9] speculated very cautiously but very presciently that the search for our remote ancestors would do best to begin in Africa. His reasoning was very simple and is worth citing in full: 'In each great region of the world the living mammals are closely related to the extinct species of the same region. It is therefore probable that Africa was formerly inhabited by extinct apes closely allied to the gorilla and chimpanzee; and as these two species are now man's nearest allies, it is somewhat more probable that our early progenitors lived on the African continent than elsewhere'.[10] He went on to allay possible criticism over any lack of evidence, pointing out that: 'the discovery of fossil remains has been a very slow and fortuitous process. Nor should it be forgotten that those regions, which are most likely to afford remains connecting man with some extinct ape-like creature, have not as yet been searched by geologists'.[11]

Although Darwin may have been the first naturalist to propose that Africa was the continent most likely to yield fossil remains of early humans, it is somewhat ironic that the first fossil evidence of ancestral humans came from Trinil near the Solo River in east Java. The fossils were discovered during 1891–92 by Eugène Dubois[12] (1858–1940), a Dutch military doctor and a skilled anatomist and geologist, who proposed the name *Anthropopithecus erectus* on the basis of a thigh bone, a tooth, and a skullcap. The modern name for this fossil is *Homo erectus*, and its most probable age is between about 0.7 and 1.0 million years ago.[13] We must go back in time in order to appreciate the full significance of this early find, beginning with a few simple but necessary definitions. The term *hominin* (previously called *hominid*) includes modern humans (*Homo sapiens*), extinct humans (such as *Homo erectus*), and our putative remote ancestors (*Australopithecus* and *Ardipithecus*).

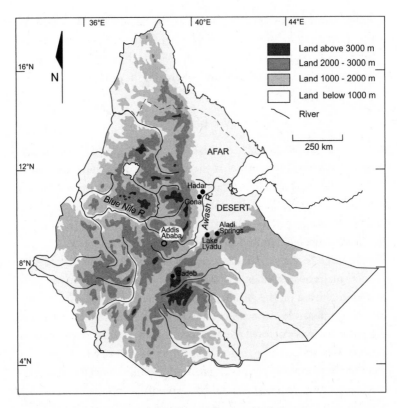

FIGURE 7.1. Map showing places cited in the text.

By far the best dated and most abundant early fossil hominins are those from the Middle Awash Valley[14] in the hot, arid western Afar Desert of Ethiopia (fig. 7.1). It is from here that the well-known partial skeleton of Lucy,[15] more formally known as *Australopithecus afarensis*, was unearthed in 1974. A very fine cast of Lucy was on display during my recent visit to the National Museum in Addis Ababa, Ethiopia, in September 2019. The original is carefully stored in a separate part of the museum. Lucy had a small brain, with a cranial capacity between about 375 and 500 cubic centimetres (cc). Modern humans have an average cranial capacity of 1400 cc, or slightly smaller than our extinct Neanderthal cousins. However, unlike the great apes such as the modern gorillas and chimpanzees, Lucy's femur or thigh bone shows that she was fully capable of walking upright. In a word, she was bipedal. The *Australopithecines* lived in Africa between 4.1 and 2.5 million years ago. They have not been found anywhere else. Their predecessors, known as *Ardipithecus*, have also been recovered from the Middle Awash area, and lived here roughly 5.8–4.3 million years ago.[16] They too are unique to Africa. These ages refer to the Middle Awash fossils, are valid as of late July 2019,[17] and will no doubt be refined as a result of ongoing research. The fossils represent decades of painstaking searching and meticulous excavation by palaeo-anthropologists like Tim White,[18] Yohannes Haile-Selassie,[19] and many others under harsh and sometimes dangerous conditions.

The late Bill Bishop once observed that certain parts of Africa were good places for hominins to live in, die in, and be found again.[20] The Afar Desert is one such place. When the Ardipithecus and Australopithecus hominins were alive, the area in which they were living was a land of perennial lakes, rivers, and wetlands, with a variety of habitats ranging from riparian forest to savanna woodland and savanna grassland.[21] Food was abundant in what was then the Afar savanna, and the climate was both milder and somewhat less arid than it is today.[22] The evidence for these statements includes fossil remains of animals and plants, especially pollen grains, fossil soils, and sediments laid down in lakes and rivers.[23] Rapid sedimentation from streams flowing down from the adjacent eastern escarpment of the Ethiopian Highlands meant rapid burial after death. Sporadic volcanic activity, always frequent in this volcanic region where three tectonic plates are moving apart,[24] meant that sediments were often buried beneath a protective cover of basaltic lava and so preserved for millions of years. Earthquakes and faults are common in the Afar, so that previously buried deposits later became exposed at the surface, awaiting the eagle eye of fossil hunters to detect the tiny and often fragmentary remains of bones and teeth.

In one locality 1400 miles/2250 kilometres northwest of the Middle Awash, a hominin fossil even older than the oldest fossils from the Afar Desert has been recovered. In the Djurab Desert of the northern Chad

Basin (map 1), near a site with an abundant savanna fossil fauna, Michel Brunet and his colleagues found the remains of a very late Miocene hominin[25] with a cranial capacity of about 360 cc. They named this hominin *Sahelanthropus tchadensis*, after its geographic location, and it lived here seven million years ago.[26] There is some evidence that this creature may have had an upright posture. They also recovered fossil remains of Australopithecus from this region.[27] Based on what has been found so far, we can safely say that the oldest African hominins found in the southern Sahara were living there seven million years ago, which is when the Sahara was becoming more arid, while the Ardipithecus and Australopithecus hominins from the Afar Desert were living there between about 6 and 2.5 million years ago. The lifestyle of these creatures was probably not too dissimilar from that of modern-day chimpanzees, with a diet of nuts, fruit, roots, tubers, and insects supplemented by opportunistic hunting of smaller animals.[28] The tools they used probably consisted of sticks, twigs, and occasional stones, much like the tools used by chimpanzees today. Once used, the tools likely were discarded. But the time was now ripe for a more sophisticated approach to tool use—one involving deliberate modification of pebbles to produce stone flakes with a sharp cutting edge.

The Earliest Stone Tools

The earliest evidence of deliberate stone toolmaking comes from the Gona valley (fig. 7.1) in the Afar Desert of Ethiopia.[29] The tools were made about 2.5–2.6 million years ago and consist of pebbles from which a few flakes have been struck using another stone. The tools lie between two beds of volcanic sediments laid down during volcanic eruptions, both of which have been accurately dated.[30] These early tool makers were adept at hitting one stone with another to split off a series of sharp flakes. A short distance from the Gona valley, on the left bank of the Awash River at Hadar (fig. 7.1), not far from where Lucy was first discovered, similar stone tools have been found together with early *Homo* fossils and have a precise age of 2.33 ± 0.07 million years.[31] The name given to this pebble tool tradition, which was also documented by Mary and Louis Leakey[32] at Olduvai Gorge in Tanzania nearly a century ago, is *Oldowan*, in honour of the Maasai name for that famous gorge.

The Oldowan tool kit consisted mainly of pebbles from which some flakes had been detached, the flakes themselves, and some stones that could be used as hammers. It may not seem very much to us today, but it did give our earliest toolmaking ancestors one big advantage. Unlike the big cats and other carnivores, humans can cool themselves by sweating and associated evaporation. And so, while the big African cats were snoozing under a shady tree in the heat of the day, our ancestors could creep out around noon with their sharp flakes and cut chunks of concentrated protein in the form of meat from the carcasses killed earlier by the ancestral lions and leopards. Discarded bones could be broken open with their hammer stones to provide marrow, a major source of fat and protein. From this time on our ancestors, who are best described as opportunistic scavengers, began to develop larger and more complex brains. The pattern of change was very slow, however, and the Oldowan tradition persisted with little change for about a million years.

The next major change in early human cultural development took place, apparently quite suddenly, at about 1.6–1.5 million years ago.[33] Our ancestors now knew how to detach large stone flakes up to fifteen centimetres (six inches) long from the parent core stone. They then struck these flakes with a stone hammer to detach pieces from around the edges on one or both sides. If flaked on one side, the tool is a uniface flake; if flaked on two sides, it is a biface flake or, more simply, a biface (plate 7.1). A biface with a pointed tip is rather misleadingly called a hand-axe, and one with a straight tip is known as a cleaver. These were multipurpose tools, the Early Stone Age equivalent of a Swiss army knife. This stone toolmaking tradition is called Acheulean or Acheulian after the small town of Saint Acheul in the north of France where the amateur archaeologist Jacques Boucher de Crèvecœur de Perthes (1788–1868) had excavated similar bifacial tools in the river gravels not far from the River Somme during the 1830s.[34] Early Stone Age bifacial tools can be found throughout the Sahara.

The Oldowan and the Acheulian together comprise the Early Stone Age or Lower Palaeolithic. The term Early Stone Age refers to Oldowan and Acheulian stone tool assemblages found south of the Sahara, and the term Lower Palaeolithic refers to those found north of the Sahara. The terms were devised when archaeologists thought that the Sahara had always been a barrier for human migration. Once Early Stone Age stone tools began appearing everywhere in the Sahara, the distinction no longer served any useful purpose, so I will use these terms interchangeably and do likewise for the terms Middle Stone Age/Middle Palaeolithic and Late Stone Age/Upper Palaeolithic.

At the upland site of Gadeb in southeast Ethiopia (fig. 7.1) where I was working with Desmond Clark and his team of archaeologists, we found both Acheulian and what Desmond Clark considered to be Developed Oldowan stone tools close together in the same localities and in the same stratigraphic horizons, showing that they were the same age.[35] We initially thought that we were looking at the tools left by two different groups of people. It soon became clear to us that what we were seeing in fact represented two quite distinct sets of activities by the same group of people,[36] with the Developed Oldowan pebble tools and associated flakes used for butchery purposes and the Acheulian tools used for other purposes, such as debarking trees to obtain the edible cambium layer beneath the bark. Other possible uses for these tools with their serrated edges might have involved skinning, cutting hides and ligaments, and sawing through bone and wood. When an undergraduate student at Cambridge once asked the distinguished prehistoric archaeologist Miles Burkitt (1890–1971) what Acheulian hand-axes were used for, he famously replied: 'I don't know; I wasn't there'. Although an honest answer, it was not very helpful. Fortunately, we have progressed quite a lot since then, and experimental studies of use wear and of plant and animal residues on the cutting edge of stone tools can now tell us quite a lot about stone tool use.[37]

The Discovery of Fire

As time progressed the Acheulian bifaces were more finely made and in outline more closely resembled the leaves of some tropical tree as if the maker had a clear mental image of the shape desired in the final tool.[38] It is at about this time that fire appears in the archaeological record,[39] as hinted at in the title of this chapter and the quotation at the chapter opening. We do not know just when humans first began to use fire deliberately, but we can guess that humans out in the East African savanna grasslands had long witnessed grass fires caused by lightning strikes during the dry season. Trees struck by lightning can burn for days, especially if growing through a termite mound. Humans living close to some of the active volcanoes in Ethiopia and Kenya would have seen fires caused by red-hot lavas. Burning branches brought into caves would have kept predators such as hyenas and jackals a safe distance away. We can further speculate that fire allowed meat and fish to be smoked and preserved from decay. Cooked meat is easier to digest, especially for infants. There are some questions in prehistoric archaeology for which we may never have definite answers. As W.H. Auden (1907–1973) concluded in his final poem *Archaeology*, 'guessing is always more fun than knowing'. What we do know, however, is that small groups of Acheulian hunter-gatherers equipped with a simple and relatively unspecialised stone tool kit of no more than about a dozen items, but with the knowledge of how to make and use fire, moved out of Africa (no doubt on many occasions) and after many generations successfully reached the limestone caves at Choukoutien near Beijing in northeast China,[40] where the winters are long and very cold. For people coming from the south, this journey meant crossing what are now the great deserts of the Sahara and its eastward extension the Sinai Desert. Those migrating north from East Africa and Ethiopia may have moved up the Nile Valley and would still have needed to cross the Sinai Desert. Did they cross when the climate was wetter, as most commentators suggest? Or could they have crossed these deserts regardless of whether it was wet or dry? In my view, both options are possible for people who were accustomed to desert life. For large groups, crossing during wetter times would have been easier, but watering points were probably always available to support smaller groups of desert nomads, much as in the Afar and Australian deserts before any roads existed.

Prehistoric Hunters of the Saharan Savanna

Louis Leakey proposed the name *Homo habilis*[41] for the first stone tool makers and claimed that *Homo erectus* succeeded *H. habilis* in East Africa about 1.5 million years ago. Present evidence shows that *H. habilis* and *H. erectus* coexisted in northern Kenya for half a million years. Two other toolmaking species (*H. ergaster* and *H. rudolfensis*) also lived in arid northern Kenya about 1.9–1.3 million years ago.[42] As more discoveries are made, the once simple human family tree begins to resemble a jungle thicket. A further complication is that some taxonomists are inveterate splitters, while others are more cautious lumpers. We also need to bear in mind that more than five million years of geologic time are represented by only a few hundred often very fragmented individual hominin fossils, which amounts to about one fossil every 25,000 years. Still, we do know that the cranial capacity of *H. erectus* and its successors increased steadily from around 775 cc at 1.6 million years ago to 1300 cc by about 0.2 million years ago.

Between about 0.5 and 0.3 million years ago we see another major technical innovation termed the *Levallois* technique,[43] which is the trademark of the Middle Stone Age. This technique involves striking off stone flakes up to 2 inches (5 cm) long parallel to the long axis of a carefully selected parent stone blank. A *Levallois core* is what remains of the parent stone blank once the desired flakes have been struck from it using a 'soft' hammer of bone or antler. The thin flakes detached in this way were then worked carefully with a soft hammer to form blades, some of which were pointed and attached with plant resin or gum to wooden shafts to form very serviceable spears. These *Mousterian*[44] points (fig. 7.2a) are characteristic of a hunting tradition. Core preparation and hafting are diagnostic of the Middle Stone Age/Middle Palaeolithic. Spears tipped with sharp stone points allow for more effective hunting. Core preparation, as described above, allows for the production of more blades per unit volume of stone.

The Middle Stone Age was also a time of regional specialisation, evident in the tanged points of the Saharan Aterian[45] tradition (fig. 7.2b), and the use of a more elaborate stone tool kit, not to mention artefacts

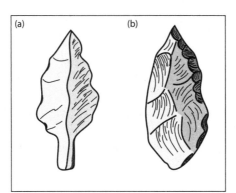

FIGURE 7.2. (a) Aterian tanged point.
(b) Mousterian point.

fashioned from wood, straw, or leather. It was during the very late Acheulian and the following Middle Stone Age that habitats in Africa that were previously mostly avoided, such as rain forests and deserts, now began to be occupied, at least intermittently. Both the Saharan Mousterian and the Saharan Aterian traditions are associated with anatomically modern humans (*Homo sapiens*), who first appear in Morocco, the Afar Desert, and the Omo Delta region of Ethiopia between 200,000 and 300,000 years ago.[46] Sites of broadly similar age with fossil remains of *H. sapiens* are also found in Israel.[47]

Another major cultural change took place in east and north Africa about 50,000–40,000 years ago. In addition to making use of a much wider range of materials such as bone, antler, ivory, and shell to make tools, this Upper Palaeolithic or Late Stone Age cultural tradition is remarkable in its later stages for rock art (paintings and engravings) and production of very finely crafted stone artefacts that were probably used for ritual exchange rather than daily use. During this stage there was a greater emphasis on seeking out high-quality sources of stone for toolmaking, and trade in both artefacts and raw materials took place over distances of hundreds of miles.[48] These exchanges would have promoted the development of social networks and reciprocal alliances to provide a form of insurance for desert dwellers in times of drought.

The Late Stone Age tool kit was diverse and specialised, with smaller and smaller stone tools (often called 'microliths') made from carefully worked parent stones to provide a much higher proportion of sharp-edged implements from increasingly smaller cores. Sickles appear in the later stages of the Late Stone Age[49] and consisted of a set of small, worked flakes with a sharp edge along one side, which would have been attached to a handle of curved wood, bone, or antler with some form of gum or resin derived from local grasses or trees. The sickles were used to harvest wild cereal grasses such as the wild species of sorghum and millet growing along the southern Sahara and of wild barley along the northern margins. Even today, the grains of many wild cereal grasses are routinely collected and eaten across the Sahara and its margins.[50] Grindstones were also common at this time and were an additional bit of evidence that these Late Stone Age folk were collecting and eating the seeds of wild grasses. Grindstones come in two parts and consist of a wide, upwardly concave lower grindstone (plate 7.2) generally fashioned from sandstone and a smaller upper grinder made from harder rock that is gently convex on one or both sides. Another innovation of these Later Stone Age people, variously called 'Mesolithic' or 'Epi-Palaeolithic', was the firing of clay to make pots used for cooking and storing grain away from rodents.[51] Sickles, grindstones, and pottery meant that the hunter-gatherers of the Sahara were already prepared for what followed—the Neolithic domestication of plants and animals, which we touched on earlier, in chapter 3.

The domestication of wild cereal grasses like wheat and barley originated about 11,000 years ago in the Fertile Crescent region of Anatolia and the upper catchments of the Tigris and Euphrates rivers as well as the Zagros Mountains of Iran.[52] Millet and sorghum were domesticated much later and most likely along the southern margins of the Sahara as well as further afield in India.[53] The domestication of cattle, sheep, goats, and pigs is again a Near East innovation. Domestication of cattle extends back to about 10,500 years ago in Anatolia, western Iran, and the Levant.[54] Earlier claims from the sites of Bir Kiseiba and Nabta Playa (map 3) in the eastern Sahara that cattle were domesticated independently from native wild aurochs no longer seem tenable.[55] Radiocarbon ages show that the first appearance of domesticated cattle at different sites in the Sahara resembles the advance of a bow wave, with ages becoming progressively younger from the Nile Valley in the east westwards across the Sahara.[56] Genetic evidence supports the view that Neolithic cattle in Africa came from a site of early domestication in the Middle Euphrates Valley and first appeared in the Nile Valley slightly more than 8000 years ago.[57] It seems that domestic sheep and goats were already present in northeast Africa before domestic cattle arrived.[58]

A Land of Green and Gold

Each of these prehistoric cultural changes took place against a background of constantly varying climate and environment. Some of these changes were slow and almost imperceptible; others were more rapid and certainly evident to the elders with their acute knowledge of how best to survive changes that had an adverse effect upon water and sources of plant and animal food. During the wetter phases, plant numbers increased, deep and shallow groundwater supplies were replenished, and the Sahara became a land of verdant savanna woodland and savanna grassland. Grasses and certain trees colonised previously active dunes, which became stable and, given enough time and sustained seasonal rainfall, developed quite deep soils. Rivers flowed all year, or at least during the rainy season. Lakes and wetlands fed by rainfall, runoff, or higher groundwater levels returned once more and were in turn home to fish as big as Nile perch (*Lates niloticus*), which can attain two metres in length, as well as crocodiles and hippos. Aquatic snails lived in the lakes and wetlands. Some of them, like the semi-aquatic *Pila*, grew to golf-ball size or larger and could be collected for bait or boiled in clay-fired pots for food.[59] Migratory birds were attracted to the swamps and lakes and could be snared for food. Reeds like *Typha* grew around the lake and swamp margins and their roots were a useful source of carbohydrate.[60] The small black seeds inside water-lily bulbs could be ground and made into porridge, much like in the Afar Desert today.

These halcyon days of reliable food and water never lasted for long. During the time when *Homo sapiens* roamed the Sahara, each full glacial cycle consisted of a relatively brief wet interglacial followed by a progressive buildup towards colder and drier conditions. By 20,000 years ago, conditions around the globe had become very tough for all warmth-loving plants and animals.[61] The Sahara was in the throes of a drought that persisted on and off for about 5000 years. With our short memories we find that hard to imagine. This long arid interval had some dramatic effects. As the trees and grasses that had once covered and stabilised earlier generations of desert dunes and sandplains died from lack of water, the dunes were reactivated and the effective southern margin of the Sahara 20,000 years ago extended between 500 and 800 kilometres/300–500 miles further south, well beyond its present southern limits, with desert dunes mobile and active in places that today receive more than 20 inches (500 mm) of rain per year.[62] I do not mean to imply that the desert dunes migrated that far south, although there was undoubtedly some southward movement of dunes. What I mean is that previously fixed dunes became mobile once again as their protective cover of plants died and the shallow soils and lichen crusts became buried beneath moving sands.

The Sahara had become a land of golden dunes where before the dunes had been carpeted in green. Grasses like *Panicum turgidum,* which have edible seeds, had now vanished from the flanks of the dunes. The mighty rivers that had once flowed across the Sahara became seasonal, then ephemeral, and finally dried up and were buried beneath the shifting sands. The former lakes dwindled, became saline, and then dried out. Strong winds blasted the brittle mud curls on the surface of the now dry lakebeds into fine fragments and blew them away as desert dust. Only the hardiest of desert dwellers remained and survived, including antelopes like the addax that are supremely well equipped for living in deserts. The animals that were not adapted to dry conditions and that were unable to leave in time died of thirst. The wind buried their bones in a shroud of sand.

The long drought finally petered out and the rains returned once more to the Sahara shortly after 15,000 years ago. The desert blossomed. Plants and animals flourished anew across the former desert. Small groups of Late Stone Age hunter-gatherers moved into the Sahara from all directions, some from the Nile valley to the east, some from the forests of West Africa, some from the northern coastlands. Lakes reappeared and rivers flowed. Fish and other aquatic animals migrated along the rivers to the many small lakes that were now scattered like confetti across the Sahara.[63] Plenitude resumed. But it was not to last. By a sad irony, the very warming across the planet that had helped to switch on the summer monsoons with their life-giving rains now had the opposite effect. A sudden influx of glacial meltwater into the North Atlantic on a gigantic scale interfered with the ocean circulation and brought very cold conditions to North America and Europe and prolonged cold and drought to North Africa. This period is known as the Younger Dryas[64] and lasted more than a thousand years, from 12,800 to 11,500 years ago. The impact of the Younger Dryas drought was like a return to the bad old days of the last glacial maximum of 20,000 years ago. Lakes and rivers dried up, dunes became mobile,[65] plants and animals mostly disappeared from the Sahara as did the Late Stone Age people, apart from some who managed to survive in sheltered mountain valleys or in oases fed by groundwater springs.

The Younger Dryas drought finally came to an end, and from about 11,500 years ago onwards, the climate became wetter. Summer insolation[66] 9000 years ago was about 6 percent greater than today in tropical northern latitudes, and the summer monsoon was both stronger and more widespread, bringing rain as far north as latitude 21°N at Selima Oasis (map 3) in the now hyper-arid eastern Sahara. The Sahara was studded with numerous small lakes, and Lake Chad deepened and expanded until it began to overflow into the Niger River and into the Atlantic.[67] At its Holocene maximum, about 9000 years ago, Lake Chad (plate 7.3) covered almost 400,000 square kilometres or almost 150,000 square miles, which is equivalent to the area of Montana, slightly greater than the area of Norway, but only 4.3 percent of the 9.2 million square kilometres of the Sahara. The grains of pollen that were blown into these lakes indicate that savanna woodland and savanna grassland were flourishing across much of the southern and eastern parts of the Sahara, attracting both animals and Mesolithic hunter-gathers. At Selima Oasis the northward displacement of the Saharo-Sahelian vegetation belt (and of the

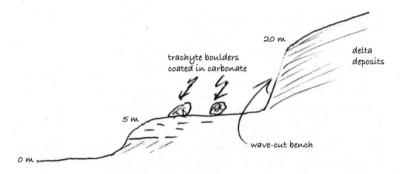

Former lake sediments, Deriba Caldera, Darfur, W Sudan (28/1/76)
(present lake shallow and very salty)

FIGURE 7.3. Sketch of a wave-cut bench eroded in older delta deposits within Deriba caldera, Jebel Marra volcano, Darfur Province, western Sudan (28 January 1976). The delta is older than the 5-metres-deep lake, which is dated between 23,000 and 19,000 years ago, and was deposited when the lake was 25 metres deep.

associated 450–100 mm summer rainfall zone) would have amounted to 400–450 kilometres between about 10,000 and 6000 years ago.[68]

The response of different parts of the desert to this general increase in precipitation would have varied from place to place, depending on geomorphic factors like relief, position in the landscape, nature of the soils or sediments, bedrock type and amount exposed, and proximity to the water table. For example, lakes fed primarily from groundwater tend to be less responsive to minor climatic fluctuations in contrast to lakes fed solely from local rainstorms and runoff. Lakes with a very large catchment area relative to their size will respond more rapidly to changes in runoff than lakes confined within small catchments. Lakes located within volcanic craters like the lakes inside Jebel Marra volcano[69] (map 1) in northwest Sudan will depend mostly upon local rainfall and so act as rain gauges (fig. 7.3).

In a classic study of the distribution of trees in relation to rainfall and soil published in 1949, the forester J.D. Smith[70] investigated the trees growing along a 1000-kilometre north-south transect across the Sudan (now two countries: Sudan and South Sudan). He found that in the drier regions of the Sudan, acacia species that were growing on sandy soils only needed about two-thirds of the precipitation required than when the same acacia species was growing on clay soils. The reason for this is that water within sand is freely available for plant roots to use, but in clay, water is held under tension and is harder for plants to access. This discovery has some interesting implications. For instance, sand laid down across preexisting clay soils during a time of windier and drier conditions would in due course be able to support a denser tree vegetation than previously even though the precipitation had diminished— a counterintuitive occurrence.

Another example of a counterintuitive response to climate change concerns sand dunes. The activity of any dune is governed by a variety of factors, including the obvious ones of strong winds and a depleted plant cover. Aridity is of course conducive to depleting the plant cover, but an important factor that is sometimes forgotten is sand supply. Dunes in many humid coastal areas are often active even when the vegetation is quite dense because the winds are very strong and there is an abundant supply of beach sand. Rivers flowing through semi-desert areas often carry a sizeable bedload of sand. If local winds blow strongly from one direction and the riverbanks are reasonably free of any dense sand-trapping vegetation, long, narrow dunes form and extend downwind (see chapter 5, fig. 5.8b). These dunes, known as 'source-bordering dunes', are not diagnostic of extreme aridity because all that they require is a regularly replenished sand supply if they are to continue to grow.[71] Seasonally flowing rivers in semi-arid areas will ferry in the necessary supply of fluvial sands for the wind to blow out of the river channel and fashion into linear dunes.

When Things Fell Apart: Migrate, Adapt, or Become Extinct?

The Sahara benefitted from a mostly humid climate between roughly 15,000/14,500 and 5000/4500 years ago, although there were periodic dry phases, of which the Younger Dryas drought between 12,800 and

11,500 years ago was probably the longest. It was during the wetter intervals that Mesolithic hunter-gatherers developed the precursors of what became the Neolithic herding and farming cultures: microliths, grindstones, pottery.[72] In some places, as in the Fayum Depression (map 3) west of the Nile, the change from hunting and gathering was almost imperceptible, with herding and/or domestic cereal crops simply added to the existing repertoire of fishing, hunting, and collecting wild plant foods.[73] In other localities such as Gobero (map 1), situated 500 kilometres south of Adrar Bous (map 1), the abundant burials in the Neolithic graves indicate the arrival of an entirely different group of people with a quite different skull morphology.[74] As we saw in chapter 3, these people were cattle herders, and the strontium isotopic composition of their teeth indicates a very different diet from the Mesolithic diet. Once again, the good times did not last. The summer rains failed to reach as far into the Sahara as they had previously. The pastures died, the lakes dried up, and most have remained dry ever since. The Neolithic pastoralists who had engraved and painted such a memorable artistic legacy across the Sahara in which were depicted lively scenes of cattle camps now faced a time of tough decisions. Should they stay and try to adapt to these new harsh conditions, and perhaps perish in the attempt, or should they move out and follow the rains further to the south? In the final chapter of this book, I will consider some of the ingenious ways in which the descendants of those few who chose to stay managed to adapt to the harsh laws of the desert. Most, however, chose to leave.

As I write these words, bushfires of unprecedented ferocity are raging across every state in my adopted land of Australia, and many of us are facing the daily challenge of whether to stay and fight or to seek safety in flight. I can therefore empathise with the dilemmas confronting those early Saharan pastoralists.

The Great Exodus

The onset of aridity in the Sahara from about 4500–4000 years ago onwards did not happen at the same time across the entire Sahara. From a climatic regime before then that modern weather forecasters might describe in terms such as 'mostly wet with dry intervals', the climate changed to 'mostly dry with moist intervals'. However, once the rivers ceased to flow and the lakes dried out, a sedentary pastoral existence was no longer possible except in upland regions with permanent water or lowland depressions where the ground surface intersected the local or regional water table, enabling plants to grow in what we now call desert oases. It was a time of Neolithic and later migrations from a number of previously occupied localities in search of reliable supplies of water, pasture, and food. Some Saharan groups moved east to the Nile;[75] others moved south to the tropical woodlands of West Africa; others sought refuge in adjacent mountains. It was the time of the great exodus portrayed in the stories and written accounts of many former Neolithic and later farming and herding communities. Sometimes the immediate cause of migration was territorial battles[76] over access to water and pasture—battles that continue to this day in many parts of Africa during times of extreme drought and breakdown of social structures designed to minimise conflict between herder and farmer, between Cain and Abel, between cowboy and sodbuster. We continue this story in the following chapter.

PART THREE

THE SAHARA TODAY

In part three we look at what the Sahara is like today. The causes and consequences of historic droughts are the subject of chapter 8, together with the fascinating but thorny problem of how extreme climatic events such as prolonged droughts have influenced human societies. In chapter 9 we explore the complex question of desertification, including the different and sometimes unhelpful ways in which it has been defined. At present, opinions are often polarised over whether desertification is a natural hazard or a result of human mismanagement. There is no sensible reason to deny that natural climatic fluctuations and human activities have played a role in causing land degradation in the arid and semi-arid world. Of relevance here is the important and much debated issue of whether human activities invariably aggravate the impact of such extreme events, or whether they can in appropriate circumstances help to minimise their impact. Chapter 10 explores how plants, animals, and human communities have adapted to living in an arid land, where rainfall is unreliable and scant, and coping with uncertainty is part of everyday life. For many hundreds of years, pastoral societies have adopted a nomadic lifestyle that allows them to make optimum use of regional fluctuations in precipitation. Unimaginative or repressive central government policies designed to re-settle nomads into fixed localities and to prevent traditional movement of nomads and their herds across what are now national borders have generated hardship, social unrest, sporadic rebellion and, more and more often, guerrilla warfare, as witnessed today in western Sudan, Chad, Niger, and Mali.

CHAPTER 8

In the Land of Great Drought

Water, water everywhere, nor any drop to drink.

SAMUEL TAYLOR COLERIDGE,
THE RIME OF THE ANCIENT MARINER

For the prophets of old, the incidence of droughts was a recurrent theme, as the following quotations from the Old Testament show so clearly: *The Lord shall make the rain of thy land powder and dust* (Deuteronomy 28:24); *I did know thee in the wilderness, in the land of great drought* (Hosea 13:5). Once the rains returned, like good news from a far country, the desert would blossom once more: *As cold waters to a thirsty soul, so is good news from a far country* (Proverbs 25:25); *The desert shall rejoice, and blossom as the rose* (Isaiah 35:1). This pattern of recurrent droughts and floods is as true of the Sahara and its borderlands as it was (and is) of the lands of the Old Testament. In this chapter we look first at the causes of droughts in the wider Saharan region and consider some of their more obvious impacts. We then consider these events in a broader global context, noting the historic significance of short-term climatic fluctuations within the region now subject to monsoonal influence.

What Is Drought?

The word drought is generally understood to mean a reduction in the amount of rain received in a particular region sustained over a number of years. But reduced rainfall is not the only factor that contributes to drought. Higher temperatures and higher moisture losses from evaporation are another. One simple definition of drought[1] is 'a regional deficiency in soil moisture which may be caused by a combination of lower than normal precipitation and higher than average evapotranspiration'. In fact, we also need to take into account the influence of drought upon river flow and groundwater. A more comprehensive definition[2] states that 'drought is generally viewed as a sustained and regionally extensive occurrence of below average natural water availability, either in the form of precipitation, river runoff or groundwater'. In addition, if soils are so severely degraded that they are no longer able to absorb falling rain, we are dealing with another type of drought which is no less important for being so insidious and which is termed *edaphic* drought (from the Greek word *edaphos* meaning soil or earth). Edaphic drought is an important cause as well as being a symptom of dryland degradation.

The following sections expand upon the influence exerted on floods and droughts by changes in the pattern of sea-surface temperatures, focussing more closely upon the semi-arid lands along the southern margin of the Sahara generally known as the Sahel.

Changes in Sea Surface Temperatures as Causes of Droughts

The moist air masses that blow in from the oceans have always been an important source of precipitation on land (fig. 8.1). Although most of the water evaporated from the ocean falls again as precipitation over the oceans, enough reaches the land to balance the amount of runoff from the continents to the oceans. Evaporation from lakes and from the soil is another source of local precipitation on land, as is evapotranspiration from trees, shrubs, and grasses. In the case of the Amazon basin, for example, much of the annual rainfall is estimated to come from local evapotranspiration from the Amazon rain forest,[3] but at current rates of forest clearing and burning, the total annual rainfall within the basin must inevitably diminish. Evaporation from Lake Victoria in Uganda is

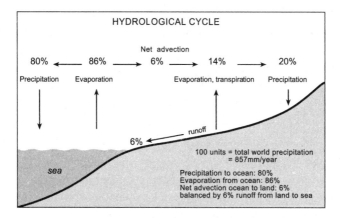

FIGURE 8.1. The global hydrological cycle. After M. Williams (2014). *Climate Change in Deserts: Past, Present and Future.* Cambridge University Press, Cambridge and New York, fig. 3.5. © Cambridge University Press. Reproduced with permission of the Licensor through PLSclear.

also considered to provide much of the annual rainfall over that vast, shallow lake.[4] In the present-day Sahara, the plant cover is sparse to non-existent, so that the Atlantic and Indian oceans and the Mediterranean Sea are the only immediate sources of moist air. How much moist air is delivered to the lands bordering the Sahara today depends to a large extent on variations from year to year in sea-surface temperatures. It is worth looking a little more closely at the causes of such variations because it will also help us understand the reasons for the recurrent droughts in the lands along the southern edge of the Sahara that are generally known as the *Sahel* (from the Arabic word for shore).

The pioneering investigations by climatologist Peter Lamb[5] in the late 1970s showed that drought in the Sahel region occurred when the zone of maximum sea-surface temperatures in the equatorial Atlantic had shifted several hundred kilometres south of its normal position. Later work by Lamb and others revealed that dry years in the Sahel coincided with times when conditions in the equatorial Atlantic were relatively warm (especially between July and September) while sea-surface temperatures were slightly cooler to the west of West Africa and to the east and north of northeast Brazil.[6] The reverse pattern was true of wetter years in the Sahel. Because the sea-surface temperature anomalies tended to persist from month to month and even from season to season, they could be used to predict whether the Sahel would receive average, below average, or above average rainfall. However, the actual amount of rainfall variance explained by the various sea-surface temperature indices was always quite limited, ranging between about 10 percent and 40 percent,[7] so that many other as yet unknown factors must also be playing a role.

In addition to the equatorial Pacific Ocean, which controls the Southern Oscillation, discussed in a later section, and the equatorial Atlantic Ocean, which has a big influence upon droughts along the southern margins of the Sahara, the Indian Ocean also has a powerful and only recently recognised influence over floods and droughts along its eastern and western sectors as well as on the Indian summer monsoon. During wetter times in the past, such as the interval between about 15,000 and 5000 years ago which we discussed in chapter 3, the moisture-bearing influence of the Indian Ocean is likely to have extended well into the eastern Sahara.

Impact of the 1968–73 Sahel Drought

The impact of the 1968–73 Sahel drought[8] is something I witnessed when I was working in the Azaouak Basin of Niger (map 1) with my geologist friend Mike Talbot, who had developed a strong interest in the geologically recent climatic history of the southern Sahara.

The Arabic word *Sahel* means shore or border and was adopted by French geographers to refer to the semi-arid southern margins of the Sahara in West Africa. The Sahel region includes parts of Burkina Faso, Mali, Niger, and Chad (map 2) and is a vague but useful umbrella term. Following two decades of above-average rainfall, the entire Sahel region experienced a severe and prolonged drop in rainfall from the late 1960s onwards.[9] This is often referred to as the 1968–73 Sahel drought, but that label is not strictly correct. The Sahel drought that began in 1968 was followed by pulses of severe drought in 1971–73, 1977, 1982–84, and 1987,

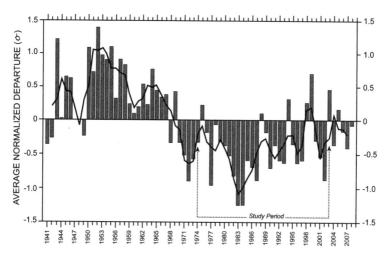

FIGURE 8.2. Time series (1941–2008) of average normalized April–October rainfall departure
for twenty stations in the West African Sudan–Sahelian zone (11°–18° N) west of 10° E.
After M. Williams (2014). *Climate Change in Deserts: Past, Present and Future*. Cambridge
University Press, Cambridge and New York, fig. 23.8. © Cambridge University Press.
Reproduced with permission of the Licensor through PLSclear.

with only minor increases in rainfall after that, followed once more by
severe drought in 2005[10] (fig. 8.2). My use of the term '1968–73 Sahel
drought' refers only to the initial five years of a series of droughts that
Michael Bell and Peter Lamb[11] described in 2006 as being 'among the
most undisputed and largest regional climate changes experienced on
the earth during the last half-century'.

In the part of Niger where Mike and I were working, the water holes
had dwindled to muddy pools of green sludge. The surrounding pastures had died off during the prolonged drought, forcing the nomadic
pastoralists who lived in this region to move south in an increasingly
hopeless search for food and water for their animals and their families.
Many of the younger men had left to find work in southern Nigeria. One
Tuareg family had set up their tent on the outskirts of the town of
Agadès (map 1) located at the southern tip of the Aïr Mountains. The
head of the family offered us tea and gave the leaves to one of his small
granddaughters to allay her hunger. All his animals had died. In the

market, sacks of millet with stencils showing they were a gift of the people of the United States were on sale in the local black market at three times the normal price. They should have been distributed for free. The wife of the President controlled the black market across the nation. I bought a sack for the Tuareg family. My host told me it would keep them alive for three months. A few weeks later I returned to the capital Niamey to catch a flight to Algiers. The airport was closed and surrounded by barbed wire and machine-gun emplacements, although I did manage to leave some days later. There had been a coup. President Hamani Diori was in prison, where he remained for ten years. His wife was shot. Government myopia and failure to acknowledge the hardships caused by drought led also to the downfall of the Emperor Haile Selassie in Ethiopia in 1974 and the fall of the Ethiopian dictator Mengistu Haile Mariam in 1991.

The Albedo Drought Model

The 1968–73 Sahel drought prompted renewed scientific interest in the causes of such droughts.[12] It also re-ignited the debate about the role of human activities in initiating or accentuating such droughts. The eminent meteorologist Jules Charney and his colleagues put forward what they termed a *biogeophysical* model to explain this drought.[13] Another name for this model is the *albedo* model of drought. Albedo is simply the proportion of incoming solar radiation reflected from the surface of the earth. Dark, moist, and well-vegetated surfaces have a low albedo; light, dry, and bare surfaces (plate 8.1) have a high albedo. Snow has a very high albedo, which is why the glare of reflected sunlight from snow can cause very painful snow-blindness. Sunlight reflected from water can be equally damaging to the eyes. So can sunlight that is reflected from dry, pale sand. One reason the Tuareg and the Bedouin cover most of their face when travelling across sandy deserts is not simply to protect their faces from the sun, but also to protect their eyes from direct and indirect sunlight.

The albedo model is simple, elegant, and seductive. It is also wrong as an explanation for the Sahel drought.[14] The model proposes that

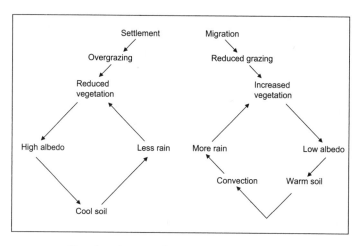

FIGURE 8.3. Hypothetical impacts of overgrazing and of reduced grazing, respectively, on plant cover, albedo (reflection from the surface), and rainfall in drylands. After M. Williams (2014). *Climate Change in Deserts: Past, Present and Future.* Cambridge University Press, Cambridge and New York, fig. 23.9.

overgrazing leads to a diminished plant cover. This in turn leads to an increase in albedo. Increased reflection of incoming sunlight causes cooling of the surface. There is less convection from a cooler surface. Less convection leads to less convectional rain. Less rain leads to a further reduction in plant cover. The predicted outcome of the model is land degradation along the desert margins and prolonged drought caused by grazing pressure (fig. 8.3). Once grazing pressure is removed and the plant cover is able to reestablish itself, the opposite occurs. Albedo is reduced, the surface becomes warmer, more convection leads to more rain and so to more plant growth. At first sight, the albedo model of drought seems reasonable and convincing. I have no doubt that on a small scale it may well apply, but it has some fundamental flaws. First, it does not explain why historic droughts begin and end at the same time in widely separated parts of the earth and in both hemispheres. Second, it does not explain why there were even more severe droughts in the past in North Africa well before there was any widespread grazing pressure.

It is useful to see the Sahel droughts against a wider and more global background because it helps to explain why, for example, times of heavy rainfall along the west coast of the United States, northwest Europe, and coastal Peru during 1982–84 coincided with times of extreme drought in other regions of the world such as the southern Sahara, India, northeast Australia, eastern China, and northeast Brazil, and why we see this pattern of events repeated again and again. In the next section we consider some of the links between global floods and droughts and the El Niño-Southern Oscillation events mentioned so often in the media, but usually without adequate explanation.

Historic Floods and Droughts: A Global Perspective

Today we are all fairly familiar with 'El Niño' events and their impact upon global floods and droughts. We may be less familiar with the term 'Southern Oscillation', although the popular media refer frequently to 'El Niño-Southern Oscillation' or ENSO events, and even talk knowingly about the 'Southern Oscillation Index'. But what do these terms really mean and what can they tell us about changes in the global patterns of floods and droughts?

To understand what is meant by the Southern Oscillation, we need to go back nearly a century to the work of Sir Gilbert Thomas Walker (1868–1958), a brilliant English mathematician and meteorologist, who was Director General of Observatories in India between 1904 and 1923. Walker quickly realised that previous attempts to predict Indian monsoonal rainfall were based on inadequate information, so he set about an ambitious programme of meteorological observations together with a search for statistical correlations with climatic events across the globe. In an important paper published in 1924,[15] shortly after he had left India, he identified what he called the Southern Oscillation. The Southern Oscillation is a measure of the difference in the surface atmospheric pressure between the western and eastern sectors of the equatorial Pacific Ocean (fig. 8.4). Walker's statistical research revealed that when the surface atmospheric pressure off the coast of Peru was below its average value, that at Jakarta in Indonesia was above its average value, and vice

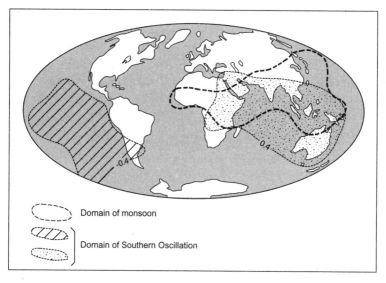

FIGURE 8.4. Region influenced by the summer monsoon and the two key regions of the Southern Oscillation. After M. Williams (2014). *Climate Change in Deserts: Past, Present and Future*. Cambridge University Press, Cambridge and New York, fig. 23.1. © Cambridge University Press. Reproduced with permission of the Licensor through PLSclear.

versa. This difference is expressed today as an index called the Southern Oscillation Index and is now usually based on the surface atmospheric pressure difference between Tahiti in the central equatorial Pacific and Darwin in tropical northern Australia. Figure 8.4 shows how most of the area influenced by the Southern Oscillation also coincides very broadly with the area subject to the summer monsoon. Changes in atmospheric circulation associated with the Southern Oscillation are known, very appropriately, as the Walker Circulation.

When the Southern Oscillation Index is strongly negative, droughts tend to occur in such widely separated regions of the earth as the highlands of Ethiopia (runoff from the highlands is the major source of Nile summer floods), the Sahel, peninsular India, eastern China, northern Thailand, Java, northeast Brazil, and eastern Australia. At the same time that these areas are experiencing drought, there is often heavy rainfall and extreme floods in other parts of the world including the southern

United States and California, coastal Peru, northern Argentina, and northwest Europe. In years when the Southern Oscillation Index is strongly positive, the opposite pattern tends to occur, with major floods common in the regions prone to drought during years of negative Southern Oscillation Index, and extreme dry conditions in California, coastal Peru, and northern Argentina. Severe bushfires often occur during the drought years, many of them triggered by dry lightning strikes.

Walker had identified the area off the coast of Peru as a critical region in his search for more reliable ways to predict the behaviour of the Indian summer monsoon. In fact, Peruvian fishermen have known for many centuries that in certain years, around the month of December, the normally cold and nutrient-rich waters offshore which support abundant stocks of anchovies and other fish are replaced by warmer water, and the fish stocks plummet, bringing great hardship to the fishing communities. Peruvian fishermen have long referred to these years as El Niño years because the warning signs usually occur around December. El Niño is Spanish for the little boy and refers to the birth of the Christ child in December.

We now have a fairly accurate and reasonably complete record of El Niño years covering the last five centuries in coastal Peru.[16] Such years do not only have a severely adverse effect on fishing; they are also years when the normally arid coastal regions of Peru are afflicted with severe floods and landslides, further adding to local misery. In El Niño years, the Southern Oscillation Index is strongly negative and the surface atmospheric pressure off the coast of Peru is lower than average. In some years the opposite situation prevails, and the waters off Peru are even colder than usual. Such years are sometimes called anti–El Niño years or, more simply, La Niña years. The expression 'El Niño-Southern Oscillation' event simply refers to a year marked by an El Niño event and a negative Southern Oscillation Index. The two phenomena (El Niño and Southern Oscillation) are closely related but are not synonymous. One refers to surface atmospheric pressure (the Southern Oscillation) and the other to sea-surface temperature anomalies (El Niño events).

There is a lake in Ecuador called *Laguna Pallcacocha* which is very sensitive to El Niño events. Sediment cores from this lake have provided a record of such events back to 12,000 years ago.[17] Following a long interval of about 5000 years during which El Niño events were rare, there was an increase in El Niño activity from about 4800 years ago until about 1600 years ago, after which there were fewer El Niño events. Frequency is of course not the same as intensity or magnitude, but I cannot help wondering whether another factor leading Neolithic pastoral nomads to move out of the Sahara between 5000 and 4500 years ago was an increase in regional climatic variability added to an overall drying trend.

Research carried out during the last thirty years or so has demonstrated that in a number of areas adjacent to our hot deserts, including southern and eastern Africa, northeast Brazil, New Mexico, eastern and northern Australia, central India, and northeast China, the incidence of wet and dry years is strongly influenced by the incidence of El Niño-Southern Oscillation events. Because the rivers in these regions are very sensitive to small changes in rainfall, El Niño-Southern Oscillation events tend to amplify their already highly variable flow regimes.[18] This tendency seems likely to persist in the future.

Even in a region as wet as Java, years of slower than average growth in teak trees (*Tectona grandis*) during the period 1852–1929 coincided with years of negative Southern Oscillation Index.[19] During the period 1877–1998, years of negative Southern Oscillation Index were years of drought and widespread fires in Indonesia.[20] It is interesting to note that years of narrow tree rings in Java were also years of below-average flow in the Nile at Cairo, indicating much reduced summer rains in the Ethiopian headwaters of that great river.

We are currently in a position to pull together a considerable body of research on the global impact of El Niño-Southern Oscillation events. We now realise that El Niño-Southern Oscillation events are a major source of rainfall variability today and were in centuries past, accounting for up to 50 percent of rainfall variance in regions as widely dispersed as northeast Brazil, India, eastern China, eastern Australia, Indonesia, Thailand, Ethiopia, southern Africa, and the southern Sahara. The regions of annual pressure anomalies associated with the Southern Oscillation extend well beyond the two tropics in both hemispheres.

However, the correlations between different localities, although statistically significant, do not account for all of the year-to-year variation

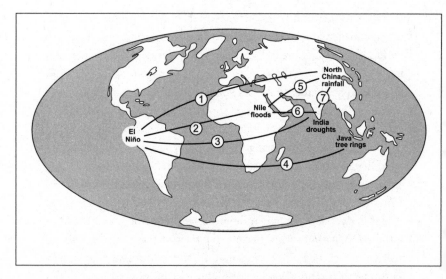

FIGURE 8.5. Statistically significant correlations between China rainfall, Indian droughts, tree rings in Java, Nile flood height and El Niño occurrences in Peru for different time intervals between 1740 and 1984. 1 is El Niño and North China rainfall 1770–1879; 2 is El Niño and Nile floods 1770–1879 and 1880–1984; 3 is El Niño and India droughts 1770–1869; 4 is El Niño and Java tree rings 1770–1984; 5 is Nile floods and North China rainfall 1870–1984; 6 is Nile floods and India droughts 1770–1887 and 1880–1984; 7 is India droughts and North China rainfall 1870–1984. Nile floods and Java tree rings correlated significantly during the 1750s and during 1870–1929. After M. Williams (2014). *Climate Change in Deserts: Past, Present and Future.* Cambridge University Press, Cambridge and New York, fig. 23.7. © Cambridge University Press. Reproduced with permission of the Licensor through PLSclear.

in the strength of the summer monsoon but do account for a modestly important part of the inter-annual rainfall variability. It is disconcerting to see that at periodic intervals, for reasons we still do not understand, the spatial pattern of variation changes quite abruptly, so that two localities that were previously in phase suddenly cease to be so. A corollary to this is that future changes in the links between floods, droughts, and El Niño-Southern Oscillation events are to be expected and may offer surprises. However, the climatic record for the last few hundred years does tell us which regions of the world have been influenced by El Niño-Southern Oscillation events and it does not seem very likely that this will change much in the future. Figure 8.5 shows what are called 'teleconnections' (or statistically significant correlations) between El Niño-Southern Oscillation events and rainfall in India and China, river flow in India and the Nile, and the width of Java teak tree rings as a measure of annual rainfall.[21]

Importance of the Indian Ocean

Precipitation along the southern margins of the Sahara comes from three main sources: the equatorial Atlantic, the southern Atlantic, and the Indian Ocean. The importance of the Indian Ocean in helping to replenish groundwater supplies in the southern Sahara has only recently been recognised.[22] At the present time, in years when the sea-surface temperatures off the Indonesian island of Sumatra are unusually low, they tend to be unusually high in the western Indian Ocean, and vice versa, a seesaw-like phenomenon known as the Indian Ocean Dipole.[23] One result of this alternating pattern of sea-surface temperatures is that drought years in Indonesia associated with colder water offshore (positive phase of the Indian Ocean Dipole) coincide with exceptionally wet years in East Africa, bringing extreme and sudden floods to the White Nile, as in 1961.[24] Likewise, drought in East Africa during years of low sea-surface temperatures offshore coincides with very wet years in Indonesia (negative phase of the Indian Ocean Dipole).

The Indian Ocean Dipole operates quite independently of the Southern Oscillation and can either enhance or neutralise the impact of El Niño-Southern Oscillation events upon the Indian summer monsoon. The Indian Ocean Dipole has quite a long reach and influences atmospheric circulation over North America and the southern Indian Ocean, affecting Australia and South Africa. The positive phase in late 2019–early 2020 brought more rain to Ethiopia, which was unusually green during my visit there in October 2019, but aggravated drought in Australia, reflected in the catastrophic bushfires in many parts of that dry continent. Because summer rainfall along the southern margins of the Sahara is in part derived from the Indian Ocean, the fluctuating sea-surface temperatures associated with the Indian Ocean Dipole have a significant influence upon the incidence of floods and droughts in that region today and likely have in the past.

Volcanic Eruptions and Droughts

Volcanic eruptions can also cause drought. The cooling impact of historic volcanic eruptions like Tambora (1855), Krakatau (1883), Agung (1963), and Pinatubo (1991) is now pretty well known, although the precise processes that bring about the cooling are complex and less easily grasped.[25] The wintry climatic repercussions of the Tambora eruption that have gone down in North American history as 'the year without a summer' also wreaked havoc on farmers across Europe, caused widespread famine in southern China and caused a deadly outbreak of cholera and typhoid in what was then Bengal and is now Bangladesh.[26] In the twelve months following the June 1991 eruption of Pinatubo volcano in the Philippines, global temperatures fell by nearly one degree Celsius. This may not seem like very much, but less well known is the drought that followed the Pinatubo eruption and that persisted for several decades in southeast Asia.[27] This drought seems to have been linked to the lowering in sea-surface temperatures that followed this eruption.

A comprehensive study of monsoon failure and droughts across Asia during the past four centuries drew upon tree ring studies from more than three hundred sites.[28] Narrow rings indicate little or no tree growth, either through lack of water or lack of sunlight. The four worst droughts were those of 1638–41, 1756–68, 1790 and 1792–96, and 1876–77 and resulted in widespread famine and loss of life. The authors of this study commented that other parts of Asia were much wetter during the time of these droughts, so that there was considerable geographical variation in the spatial patterns of rainfall across the continent, much as there is today.

There are some close parallels between the Asian droughts recorded in this tree ring study and the incidence of extreme droughts and associated famines in the Sahara and its bordering regions. American climatologist Sharon Nicholson used archival evidence to compile a detailed record of wet and dry phases in North Africa during the last five hundred years.[29] She found records of a very severe Sahel drought during 1736–58 and of an extreme drought in the Senegal basin during 1749. Extreme drought in the Senegambia region in 1640 was followed by frequent droughts in the 1700s until about 1775. The region around Lake Chad suffered prolonged famines linked to drought after 1790. Algeria experienced severe famines during the 1700s. The Sahara also experienced drier climatic conditions from the early 1820s until the 1860s and an especially severe drought between 1828 and 1839 which began in the 1790s.

There is also something I find very intriguing which is not discussed in the Asian tree ring study. Each of these major drought episodes also coincides very broadly with times when the volcanic dust veil index calculated by the renowned English climatologist Hubert Lamb[30] (1913–1997) was high in northern middle and high latitudes. Lamb had originally trained as a mathematician and spent many years examining the statistical relations between historic volcanic eruptions of known magnitude and any unusual changes in atmospheric circulation, near surface temperature and net radiation flux. Using the 1883 Krakatau eruption as his standard, he then compiled a series of volcanic dust veil indices. The historic eruptions of Tambora, Krakatau, Agung, and Pinatubo have shown very clearly that the climatic impact of volcanic eruptions in low latitudes like Indonesia and the Philippines can be felt very widely across the globe.

However, recent work has also shown that eruptions in high latitudes can have a major impact upon the tropics. For example, the 1783–84 eruption of Laki volcano in Iceland was associated with substantial (−1 to −3°C) Northern Hemisphere cooling in the summer of 1783, weakening of the African and Indian monsoon circulation, greatly reduced Nile flow, and reduced rainfall south of the Sahara.[31] It also seems likely that very low Nile flow in 939 CE was associated with the eruption of Eldgjá volcano in Iceland—the largest high-latitude eruption of the last 1500 years.[32] Closer to the present, the 1982 eruption of El Chichón volcano in Mexico[33] may have aggravated the already severe drought in the Sahel. Just how these eruptions manage to weaken the regional hydrological cycle is still not fully understood, but changes in ocean circulation and sea-surface temperature undoubtedly play a major role. There is some evidence of a causal relationship between historic volcanic eruptions and El Niño events,[34] resulting in very widespread changes in sea-surface temperatures and in associated changes in rainfall leading to drought.

Recurrent Droughts and Their Global Impacts

During the years between 1877 and 1983, there were extreme and synchronous droughts[35] in northeast China, India, Ethiopia, and southeast Australia in the years 1877, 1899, 1902, 1941, 1965–66, 1972, and 1982–83. Each of these drought years were years when the Southern Oscillation Index was strongly negative, and many of them coincide with times of drought and famine in the Sahel. The impact of many of these droughts was ferocious. For example, ten million people died of starvation in China during the 1877 drought, as did five million people in India and nearly half the population of Ethiopia. Equally sadly, the return of wetter conditions was often a very mixed blessing, with sudden floods causing loss of life and livelihood, as pneumonia killed many of the already weakened people and the health of their animals also suffered from the waterlogged conditions. As a consequence, the extreme floods[36] that afflicted this vast region during the years 1887, 1889–90, 1894, 1916–18, 1955–56, 1964, and 1975—all years of strongly positive Southern Oscillation Index—were not always seen as a welcome relief, at least in the short run. They also caused soil loss through accelerated erosion. Once the fertile topsoil is lost, crop and pasture yields will inevitably diminish. I discuss this question of the causes and consequences of what is widely and inaccurately called desertification in the next chapter, where we look at the thorny topic of desertification and explore why this apparently simple and innocuous word has generated more heat than light and has led to some misguided policies.

CHAPTER 9

Human Impact on the Sahara

Dust on the waterless plains blows over his track,
The sun glares down on the stones and the stones glare back.

DOUGLAS STEWART, *THE BIRDSVILLE TRACK*

Desertification: Human Impact or Climate Change?

Humans have a curious tendency either to blame all our current environmental woes upon human mismanagement or to deny that there are any problems, and if there are, that we cannot possibly have caused them. Both of these extreme views are unhelpful. What we need most is a clear-sighted appraisal of any particular environmental issue, such as salt accumulation in previously forested soils cleared for agriculture, or accelerated soil erosion caused by a decrease in protective plant and litter cover and an increase in runoff. This should be followed by realistic and practical measures to deal with the problem in a manner that will endure and lead to no additional adverse side effects.

In the case of the Sahara (and this is equally true of our other great deserts), there is an oddly myopic view that humans were responsible for creating the desert or at least for creating deserts along and beyond the semi-arid margins, in places once verdant and well able to support plant and animal life. Two examples will suffice to illustrate this recurrent, remarkably persistent, and sadly deluded notion.

I will start with an instance that does not concern the Sahara directly but deals with the desert known in northwest India as the Thar or Rajasthan desert and in eastern Pakistan as the Cholistan desert. In their popular 1977 book *Climates of Hunger*, climatologist Reid Bryson and science writer Thomas Murray argue that this desert originated 4000 years ago entirely because of human mismanagement associated with the rise of the Indus Valley Culture. The farming communities around the towns of Harappa and Mohenjo-Daro, so the story goes, cleared the original vegetation and generated so much dust that the summer monsoon was drastically weakened, leading to sustained drought and widespread rampant desertification. In the words of Bryson and Murray:[1] 'They helped make a dustbowl out of a bread-basket and have kept it that way.'

In fact, desert dunes have been active intermittently in this region for at least 200,000 years,[2] for reasons that have nothing to do with humans but have everything to do with astronomically controlled changes in the amount of insolation received in these latitudes. The actual inception of the Thar Desert most likely dates back even further in time, to the onset of global cooling and climatic desiccation that began about seven million years ago,[3] and which is when the region now known as the Sahara also began to dry out.

The second example is one I have already cited at the start of chapter 4 and comes from Paul and Anne Ehrlich's widely acclaimed book *Population, Resources, Environment: Issues in Human Ecology.*[4] It is worth quoting once more because it contains two major and persistent fallacies: 'The vast Sahara desert itself is in part man made, the result of overgrazing, faulty irrigation, and deforestation, combined with natural climatic changes. Today the Sahara is advancing southward on a broad front at a rate of several miles per year.'

It is easy to refute both of these claims, but the thinking patterns behind them are less easy to dislodge. The Sahara came into being roughly seven million years ago,[5] or long before the emergence of Neolithic farmers and herders. Since then it has undergone progressive desiccation interrupted by wetter climatic intervals. The Sahara is dry today for good and sufficient geographical reasons that have nothing to do with humans. The two key factors governing Saharan aridity are its location in latitudes characterised by dry subsiding air, accentuated by its enormous size, so that rain-bearing winds blowing from the Atlantic Ocean, Mediterranean Sea, and Indian Ocean have shed most of their moisture by the time they reach the desert.

Along the seasonally wet borderlands flanking the Sahara to the north and south, periodic prolonged droughts have made life precarious for both farmers and herders during the last few centuries, and this situation has indeed been exacerbated by inappropriate national and international policies, as Diana Davis[6] has so eloquently demonstrated for northwest Africa in *Resurrecting the Granary of Rome* (2007) and more widely in *The Arid Lands: History, Power, Knowledge* (2016). In particular, ill-advised and draconian programmes designed to re-settle nomadic pastoralists and to substitute plantation monoculture for previously diverse native vegetation have caused far more harm than good, and have led to social unrest, distrust of the central government and, in a number of cases, local warfare.

The notion that the Sahara is advancing a few miles southwards along a broad front is one that seems to pop up about every thirty years. Its latest version stems from an aerial survey by plant ecologist Hugh Lamprey[7] of the vegetation in northern Sudan in 1975. The results of his work have been taken out of context and used by alarmists ever since. It is worth noting what Lamprey did and did not do. Between 21 October and 10 November 1975, he carried out an aerial survey of the plant cover growing along the desert margin of northern Sudan. He concluded in his report[8] that when compared to the northern limit of semi-desert vegetation in northern Sudan shown on the small-scale 1958 vegetation map of Sudan published seventeen years earlier by Harrison and Jackson,[9] the desert boundary appeared to have shifted south by roughly 90–100 kilometres or 55–60 miles during the subsequent 17 years. This would equate to an average rate of about 6 kilometres or 3.7 miles per year. It is important to realise that the 1950s were wetter than normal in northern Sudan and that Lamprey's aerial survey followed a decade of below-average rain.[10] More recent work has revealed that the plant cover along the southern Sahara varies greatly from year to year, and that recurrent satellite observations of changes in the plant cover are needed over long intervals of time, certainly more than a decade, to demonstrate any genuine trends in space and time.[11]

Historic intervals of above- or below-average rainfall in the semi-arid lands immediately south of the Sahara appear to occur about every thirty years. Intervals of prolonged drought tend to bring forth the prophets of doom. For example, the forester E.P. Stebbing,[12] who had worked mainly in India and only for a short time in Africa, sounded his warning bell repeatedly and stridently during the Sahelian drought years of the 1930s with a series of publications adorned with such apocalyptic titles as 'the encroaching Sahara' and 'the man-made desert in Africa'. Silence returned once the drought had broken and the rains resumed. What many outside observers seem unable to appreciate is that a high degree of climatic variability is entirely normal in the semi-arid world and has nothing to do with human endeavours. This brings us to the thorny question of just what is desertification?

What Is Desertification?

The great Roman historian Tacitus (AD 56–AD 120) refers to the scorched earth policy of the Roman army in his magisterial history *Agricola* (AD 98)[13] and quotes one of Rome's great enemies' contemptuous assessment of this policy: *Ubi solitudinem faciunt, pacem appellant*, or 'wherever they create a wasteland they call it peace' (my translation). Our word desert comes from the Latin word *desertus* meaning empty, deserted, abandoned. From the start of the Industrial Revolution in about 1750 onwards, there was a progressive movement of rural people from the western European countryside to the towns seeking work in the burgeoning factories—William Blake's *dark, satanic mills*. French geographers referred to this movement from the French countryside as the *desertification* or abandonment of the rural farmlands. Later the term 'desertification' was applied to land in the former French colony of Algeria that was deemed to have once been forested and fertile but was now considered impoverished as a result of the allegedly feckless land use practices of the original inhabitants.[14] Very little persuasive evidence was ever produced to back up these assertions, but they did serve as an excuse for appropriating land and dispossessing the previous owners.

In the words of one distinguished geographer, Professor Jack Mabbutt,[15] desertification is simply *a change to more desertic condition*. I have never found this definition very useful because you then need to ask yourself 'what is desertic?' However, he did point out some of the results of such a change, including impoverishment of ecosystems, accelerated soil degradation, reduced plant and animal productivity, and impoverishment of dependent human livelihood systems. When there was a combination of climatic stress and land use pressure, the almost inevitable result was land degradation, social distress and, in extreme cases, famine.

Trying to define desertification is a bit like trying to handle an oiled eel with silk gloves—elusive and slippery. It is much simpler, and I think ultimately far more useful, to regard desertification as a form of land degradation reflected in a variety of consequences. The great American conservation scientist Aldo Leopold (1887–1948) put it very well and very succinctly in *A Sand County Almanac* (1949) when he commented: *'The effort to control the health of land has not been very successful. It is now generally understood that when soil loses fertility, or washes away faster than it forms, and when water systems exhibit abnormal floods and shortages, the land is sick'.*[16]

Land degradation is a general term for a reduction in the productive capacity of the land. It can be caused by human activities and by climate change. Climate change can involve long-term desiccation as in Australia and California over the past few decades, resulting in unprecedented bushfires. It can also involve shorter term effects from drought or severe floods. Desertification is a general term for land degradation in the drier regions of the world located beyond the margins of hyper-arid deserts. Deserts are already deserts and so cannot be prone to desertification. They are deserts because they occur in latitudes that do not receive much rain.

The reasons the land is sick vary from place to place. Some of the more obvious symptoms of this sickness, otherwise known as land degradation, include accelerated erosion by wind and water, by which

I mean at a rate significantly faster than the long-term rate of denudation in that region and, more crucially, far faster than the local rates of soil formation. Another symptom that is easy to overlook is a decline in soil structural stability, which then leads to an increase in surface crusting and surface runoff and a reduction in soil infiltration capacity and soil moisture storage.

The decline in the stability of soil structural aggregates is caused by a decline in soil organic matter and soil nutrient status with an attendant decline in crop and fodder yields. In extreme cases, this loss in soil fertility and drastically curtailed fodder and crop yields can cause social disruption, famine, and human and livestock migration into more fertile areas, which can culminate in conflict with the inhabitants of those areas.

An insidious aspect of land degradation in arid and semi-arid regions along the tropical southern and temperate northern margins of the Sahara is the slow but progressive accumulation of salt in the surface and subsurface horizons of dryland soils. Leaching of salt from these soils can then lead to an increase in the salt content of previously freshwater lakes, wetlands, and rivers.

Over time, the original vegetation cover of forest or woodland is replaced by secondary savanna grassland or desert scrub. There are several reasons for this, of which one involves forest clearing and burning by farmers eager for more land. Another less obvious reason is the reduction in soil moisture storage mentioned earlier, which in turn is caused by the reduction in soil infiltration capacity or the ability of the soil to absorb rainfall.

Soils under forest will store more water than soils under desert scrub. This water is released slowly to the streams in the form of base flow, which is simply water flowing below the surface as opposed to surface runoff or overland flow. The result of increased runoff and reduced base flow is an increase in the flow variability of dryland rivers and streams, such that once perennial and reliable sources of water become seasonal or ephemeral. In addition, a change from forest or woodland to desert scrub and other dryland ecosystems will often lead to a reduction in species diversity and in total plant biomass.

Savanna grasslands are vulnerable to fire, whether caused by lightning or humans. Such biomass burning will release carbon particles and trace gases into the atmosphere and, together with the increase in dust particles blown up from the dry soil surface, can cause breathing problems, especially in very young children.

If these are some of the many consequences of desertification, what are the likely causes?

Causes of Desertification

As always when considering the natural world, we need to be careful not to rely on over-simplified generalisations. Each example of desertification must be considered on its own merits before a reasonably credible diagnosis can be made about possible causes. Simply blaming overgrazing by the nomadic pastoralists as 'the cause' of desertification is neither credible, useful, nor helpful; in fact, it is a poor excuse and an evasive substitute for clear-sighted, well-informed evaluation. Nomadic pastoralists have been successfully feeding their animals and themselves for over 5000 years in the drier parts of Africa, including the Sahara and its semi-arid borderlands. The secret of their success? An acute knowledge of the plants, of rainfall patterns, and of water holes and near-surface water supplies. Mobility, flexibility and accumulated ecological knowledge have enabled the nomadic Tuareg of the Hoggar and Aïr, the Tibu/ Goran of Tibesti and Ennedi, the Peul/Fulani of southern Niger and northern Nigeria, and the Beja/Hadendowa nomads of the Red Sea Hills (map 1) to adjust to their harsh environments and to live in relative harmony within these often fragile habitats.[17]

Another factor contributing to the pastoralists' success is the relationship they developed with sedentary cultivators. For example, before the outbreak of civil war in western Sudan, the Fur cultivators of Darfur allowed the nomadic herders from the north to graze their animals on the stubble left in their fields once the crops had been harvested. Both parties benefitted: the cultivators, from animal fertiliser on their fields, the nomads, from reliable dry season fodder for their herds. Any disputes were quickly settled by the respective and respected tribal chiefs of the farming and herding communities. This long-established way of resolving disputes before they became dangerous broke down when the central government removed power from the tribal chiefs and then proved unable, or unwilling, to provide an alternative way of peacefully resolving disputes over access to land and water. The outcome—a civil war that continues to this day.

A by-product of this war is neglect of traditional soil and water conservation measures, concentration of people forced off their land into refugee camps, stripping of the surrounding vegetation for fuel and shelter, dust storms, dune re-activation and other forms of land degradation, including once friable and fertile soils converted to bare, hard surfaces that are no longer able to absorb rain or to allow plant seeds to germinate. I use this example, which can be multiplied many times across the region, to show that political and social factors can be just as important as climatic fluctuations in causing widespread land degradation in the semi-arid lands of the southern Sahara.

Needless to say, there are multiple examples of dryland degradation for which humans were unwittingly responsible. Clearing trees to provide more land for cultivation in the Ethiopian Highlands has led to accelerated soil erosion in the Nile headwaters and accelerated siltation of reservoirs downstream in Sudan and Egypt. Over-irrigation and poorly designed drainage systems have led to waterlogging and salt accumulation in the soils of arid Egypt and Sudan. Cutting down trees growing on previously stable sand dunes has triggered a wave of dune re-activation along the semi-arid margins of the Sahara, some of it only evident in the last few decades.

Natural Desertification

It is easy to forget that there are many examples of desertification that have nothing to do with human intervention. Consider, for example, the deeply entrenched ephemeral streams in the semi-arid Matmata Hills of Tunisia (map 1) and the rugged uplands of the arid Sinai Desert (map 1) and the arid southern Negev Desert of Israel. In each case, the stream incision and ensuing gully erosion are caused by changes in sediment influx from wind and water allied to local changes in the amount and intensity of surface runoff. In the normal course of events, when the hillslopes are bare, runoff can be both rapid and intense, especially during sudden, extreme downpours. The result is a flash flood, brief but highly erosive. Such floods can start quite literally with a roar from a tributary valley, followed by a wall of water capable of shifting loose blocks of rock three feet or more in diameter and dumping them up to a mile downstream.

In the arid Alashan Plateau of Inner Mongolia in northern China, I have seen granite boulders the size of soccer balls abruptly caught up and rolled along in a sudden flood that originated in the high mountains upstream. Unlike the short-lived flash floods to which I referred earlier, this particular flood in early August 1993 was a drought-breaker. The rain lasted thirty hours. More than twelve inches of rain (300 millimetres) fell, causing local wells along the mountain slope to overflow, converting valley floors to quicksands that were tricky to navigate, and above all bringing to an end three years of drought and nearly eight years of below-average rain. The Mongolian camel herder and his wife in whose house our small party had been offered shelter told me while we were on an early morning walk next day that this rain would sustain his herding community for the next three years. He also pointed out the overflowing wells that were often hard to spot in the enshrouding dense mist. In effect, this unusually heavy rainstorm had brought to an end the desertification that had been under way in this region in the preceding eight years, when the rainfall was well below average.

Given even a modest amount of rain, bare rocky hillslopes can generate a lot of runoff. Such was the situation in much of North Africa until 30,000 or more years ago when wind-blown desert dust began to accumulate across the landscape, blanketing the rocky slopes in a mantle of silt. As rain fell, it was mostly absorbed by the permeable dust mantle. Some of the rain that fell did generate runoff which in turn carried some of the silt downslope to accumulate along the valley bottom (fig. 9.1a). Sluggish streams choked with this fine sediment flowed along the valley floors and eventually built up a layer of fine alluvial silt 30–60 feet/10–20 metres thick (fig. 9.1b). The supply of wind-blown dust to

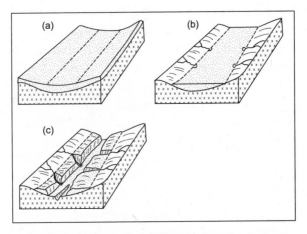

FIGURE 9.1. Natural desertification. (a) Desert dust is deposited across
the landscape and is washed downslope into the valley bottom.
(b) A thick layer of desert dust accumulates along the valley floor,
leaving the rocky hillslopes bare. (c) Intense runoff from the rocky
hillslopes causes a wave of gully erosion along the valley bottom.

the landscape slackened off about 15,000 years ago. The rainfall regime
became more intense, scouring the hillslopes of their sponge-like cover
of permeable silt and exposing the underlying impermeable bedrock.
The increase in hillslope runoff ushered in a new wave of erosion and
stream channel entrenchment. Gullies developed within the former
fine-grained valley fills and extensive gullied areas reminiscent of the
Dakota Badlands soon developed (fig. 9.1c). None of this gully erosion
had anything to do with humans, but to visitors ignorant of the recent
geomorphic history, it is all too easy to blame the local people. A great
deal of this process of natural desertification—the term used by geologist Yoav Avni[18] from his many years of study in the Negev Desert—has
been unfairly attributed to overgrazing by nomadic desert people.

When the Romans first came to North Africa, the coastal streams
were busily engaged in incising their channels, again for reasons that had
nothing to do with any human activities. As they moved upstream along
the river valleys, the Romans built a series of stone dams.[19] The aim was
to trap sediment behind the dams in order to grow olive and date palm

trees, or even wheat and barley if there was a large enough area of alluvium. The remains of these Roman dams are now left high and dry along the edges of the seasonal stream channels (plate 9.1), which have continued to cut down at intervals during the last 2000 years. The same technique of using porous stone dams to trap silt, which in turn traps soil moisture, is used to this day in Tunisia to create groves of olive trees and date palms (plate 9.2).

International Responses to Desertification

In a widely cited and influential book published in 1949 on tropical forests in Africa by the French forester Aubréville, he used the term 'desertification' to describe the impact of biomass burning on the forests he had studied and cherished for many years. He concluded that indiscriminate burning was rapidly destroying the original forest cover, creating desert-like conditions even in areas with a mean annual rainfall of 750–1500 millimetres a year.[20] He was not the first to use the term desertification for land degradation in Africa, but he is often mentioned as being the first.

A number of influential geographers in Europe and North America took up the cudgels on his behalf and began to look for evidence of encroaching desertification well beyond the limits of the present-day deserts, especially in North Africa.[21] The wish was father to the thought and they duly found the evidence they were already convinced was there to be found. It became a case of 'don't confuse me with facts; my mind is made up'. What Aubréville observed in the wetter parts of former French West Africa was savanna grassland replacing forest as a result of deliberate clearing and burning by shifting cultivators. This slash-and-burn form of shifting cultivation by tropical peasant farmers across the globe worked well when the land once cultivated was abandoned for long enough to allow trees and shrubs to regenerate. But once the human population had increased and the intervals of fallow had become increasingly more short-lived, the land suffered.[22] With the surface layer of litter and vegetation destroyed by fire, the soil was no longer protected from the erosive impact of falling raindrops and surface runoff at the start of the wet season. My own field experiments monitoring erosion in tropical northern Australia have shown that even on gentle slopes, soil loss from bare surfaces at the onset of the wet season can be many times greater than later in the wet season once a protective cover of tall grasses and herbs has been re-established.[23]

It is easy to overlook the fact that savanna grasslands were common in Africa and elsewhere millions of years before ancestral humans first began to use fire.[24] French and British colonial policy makers influenced their international counterparts with what were no doubt sincerely held views. Henceforth, desertification was regarded by the United Nations Environment Programme[25] as 'land degradation in arid, semi-arid and dry sub-humid areas resulting from *adverse human impact*' (my emphasis). This was convenient, even reassuring, because if humans were to blame, something could presumably be done about it, with international agencies advising national governments on what to do. The 1968–73 Sahel drought put a stop to this all too complacent and misguided approach to land degradation. The African nations were acutely aware that the drought was the prime cause of land degradation. The actions of desperate farmers and herders were only exacerbating the impact of too little rain. Plants died, topsoil was blown away, dunes became mobile, animals died, famine stalked the land.

During the 'Earth Summit' in Rio de Janeiro in 1992, the document called Agenda 21 was prepared and desertification was now accepted as land degradation caused by both climatic fluctuations and human actions.[26] The International Convention to Combat Desertification with a strong emphasis on Africa came into effect in 1994.[27] Individual countries were encouraged to produce Action Plans. Many did. Just how useful they were is moot. How do you assess the extent and severity of different forms of land degradation when you have no reliable quantitative observations, no trained field staff, and no means to support them? Shelter belts of trees became fashionable. Fine, except that in some parts of North Africa the winds were blowing the dunes away from the shelter belts, which were therefore of little use. The same is true for attempts to halt gully erosion and arroyo expansion when these are part of a natural cycle of erosion. As the eminent American geographer Yi Fu Tuan pointed out years ago, any conservation measures attempted in such a context will be going against nature and will probably fail.[28] The English philosopher and essay writer Francis Bacon (1561–1626) put what I am trying to say very succinctly back in 1620 when he observed that *Natura non nisi parendo vincitur*, which I translate somewhat loosely as 'you cannot manage Nature unless you first obey her laws'.

Variable Response to Climate Change

After the long dry interval between about 24,000 and 18,000 years ago that coincided with the last time North America and much of Europe were covered in ice sheets up to three miles thick, the global climate became warmer. The ice caps melted, sea levels rose, flooding the continental shelves, forcing coastal dwellers to move inland. Tropical rain forests expanded well beyond their former limits and the hot tropical deserts retreated and were replaced by savanna grassland and forest. Rivers that had dried up began to flow once more across the Sahara, and lakes re-appeared. The popular term for this wet period in North Africa is the African Humid Period, which, on the basis of a single marine sediment core collected off the coast of Mauritania by marine geologist Peter deMenocal and his team, was bracketed between about 14,800 and 5500 years ago.[29] The cutoff points were determined on the basis of a sudden decline in wind-blown dust 14,800 years ago and a sudden increase in wind-blown dust 5500 years ago. The interval of low dust flux was interpreted as humid, which seems perfectly reasonable.

Needless to say, the term African Humid Period certainly does not apply to Africa as a whole. During this same time interval, the climate in some regions of Africa was drier than today, some regions were indeed wetter than now, but in others the climate was much as it is at present.[30] It is also worth noting that as far as the Sahara itself is concerned, the onset of wet conditions was far from synchronous, and the same is true of the onset of dry conditions. The onset and the cessation of wetter conditions in the northern Sahara were not in phase with those in the south, and there were major regional variations between west and east, depending upon the source of precipitation.[31] Finally, and most important, the response of different elements of the landscape to climate change was highly variable even within the one small region.

I will illustrate this last point by referring to some excellent recent work in the southern Sahara. Lake Yoa[32] is a small and highly saline lake at Ounianga/Wanyanga in Chad (map 3). It is located halfway between Tibesti volcano and the Ennedi sandstone plateau (map 3) and today lies in the path of the NE Trade Winds that are steadily blowing sand into the lake. At Lake Yoa, geographer Stefan Kröpelin from Cologne and his multi-national team obtained two sediment cores from the upper 7.47 metres of the lake deposit. The lake silts were finely layered and contained wind-blown and aquatic pollen as well as diatoms, ostracods, and cladocera (water fleas). Diatoms are single-celled algae with an outer layer of silica (glass). Ostracods are tiny crustaceans a few millimetres in size with a calcareous shell. All three organisms are sensitive indicators of water depth, temperature, and salinity. The fossil pollen record from Lake Yoa showed that a change from savanna to semi-desert vegetation began 5600 years ago and accelerated after 4800 years ago when tropical trees vanished. Until 4300 years ago, the surrounding country was savanna grassland with some scattered *Acacia* trees. Rivers flowing down from Tibesti also brought in pollen from montane shrubs such as the tree heath or giant heather *Erica arborea*, but this ceased by 4300 years ago, indicating that rivers no longer flowed into Lake Yoa from Tibesti. Pollen from desert plants first appeared about 2700 years ago at the same time that sand was blown into the lake from the north. The sand influx seems to mark the onset of the modern NE Trade Wind regime, with the winds funnelled south between the two mountainous masses of Ennedi and Tibesti. Taking all the evidence together, the overall conclusion for this region is that desiccation was gradual, not sudden. Just as critical for any human communities, different ecosystems, whether terrestrial or aquatic, responded at different times and in different ways. In other words, we cannot ignore the influence exerted by local factors such as topography, soils, and runoff in filtering the response of local ecosystems to regional changes in climate.

Desert lakes will not all respond to climate change in the same way. Some lakes are highly sensitive to even slight changes in rainfall and runoff and are sometimes known as amplifier lakes. For example, a small lake with a huge catchment area will respond more rapidly to a change in runoff than a large lake with a relatively restricted catchment area.[33] Lakes that are fed primarily from groundwater will often be slow to respond to precipitation changes and will persist for longer in the landscape during dry climatic intervals than lakes dependent solely upon runoff following rainfall. Some Saharan lakes were fed by a combination of groundwater input and local runoff. Another factor that is easily forgotten is that some of the larger lakes derive their water from well beyond the desert. For example, Lake Chad (map 1) receives much of its water from the Adamawa Highlands in Cameroon (map 1), so that its past fluctuations were more in response to runoff changes in its equatorial headwaters than to climatic changes in the immediate vicinity of the lake. When Lake Chad was a closed basin, it was vastly more sensitive to changes in evaporation and precipitation within its catchment than when it filled its basin and overflowed to the sea.[34] At this point it became a reservoir lake, or simply one place of expanded water storage in a through-flowing drainage system. Reservoir lakes are insensitive to local climate change and simply fluctuate in response to runoff changes in their headwaters.

It is worth remembering that the Sahara is huge and covers about one-third of the continent of Africa. Its present northern limit is roughly 30° N. In the far northwest it abuts onto the southern slopes of the Atlas Mountains. Its present southern limit is about 16° N. The amount of solar radiation received in these two latitudes has varied over time, and the times of maximum and minimum insolation were not synchronous in these latitudes. Because the insolation exerts a powerful influence upon summer monsoon strength and upon convectional rainstorms more generally, we would expect to find that the onset and close of wet and dry climatic intervals will be out of phase in different Saharan latitudes. In a word, climate change will be *time-transgressive* or *diachronous*, as will be the response of rivers, lakes, and desert ecosystems.

Did Humans Cause the Sahara to Become Dry?

Archaeologist David K. Wright[35] has argued that humans were indeed effective agents in what he called 'the termination of the African Humid Period'. Harking back to chapter 3, I think we can accept that the interval between about 15,000 and 5000 years ago in the Sahara can be taken as generally more humid than today, with dry intervals, notably between 12,800 and 11,500 years ago and at 8200 years ago, with the following 5000 years generally dry with moist intervals. The question then arises, did humans speed up the climatic desiccation that was under way in the Sahara about 5000 years ago? They obviously did not cause it, because the onset of arid conditions about this time shows broadly similar timing in many other parts of the tropical world, including India and China.[36] The models cited by David Wright point to feedback effects on atmospheric circulation associated with more dust input into the atmosphere from cattle trampling and albedo changes caused by overgrazing from herds of domesticated cattle. Climate models can be very useful means of exploring likely scenarios of past and future change. They are even more useful when supported by careful long-term observations. Such observations are available for the present but are sadly lacking for the past. Even if pollen evidence shows that cattle appear at a certain site in the Sahara at the time when the evidence from fossil pollen shows a change to a plant cover more adapted to aridity, does this mean that one caused the other? If so, which?

I have long pondered this same question and can think of only two possible examples of Neolithic herding causing accelerated and locally irreversible erosion and land degradation in two desert regions in North Africa. I have worked for some months in both localities. At Adrar Bous (map 1) in the central Sahara, which we visited in the prologue to this volume, there seems to be an order of magnitude increase in the sedimentation rates during the Neolithic relative to the preceding early Holocene.[37] The increase in sedimentation implies an increase in erosion from the valley slopes. Was it caused by overgrazing by herds of Neolithic cattle? Or was it caused by a reduction in plant cover associated with a decrease in precipitation that left the sparsely vegetated valley sides vulnerable to rapid erosion from sporadic intense rainstorms? I cannot say. Perhaps it was a bit of both.

A more convincing example comes from the middle Awash valley in Ethiopia's arid Afar Desert (map 1). In this area there is a thick sequence of horizontal fluvial and lacustrine sediments older than about 10,000 years. Gully erosion is active in this area today and has reduced much of the area to a typical 'badlands' topography.[38] This gully erosion is a recent phenomenon. I was never able to find evidence of infilled gullies of similar size and extent in any of the older sediments, which were always very well exposed. Two possible causes apart from human impact are tectonic activity and climatic change. Neither agent can explain the absence of similar gully erosion in the older sediments. And so, almost by default, the only explanation that I find at all satisfactory is human impact. Just when this impact began, I cannot say. It was at some time in the last 10,000 years, certainly, and most likely not before about 4000 years ago, when herds of cattle appeared in this area at about the time that the climate became dry.

Our answer to the question of whether humans caused the Sahara to become dry is a fairly resounding 'No!' Humans may have helped to aggravate the environmental impacts of the latest phase of climatic desiccation which set in between about 5000 and 4000 years ago, but this impact was quite localised and probably caused by overgrazing from herds of domestic cattle. Prehistoric humans did not have an impact on the regional climates of the Sahara. On the contrary, it was the changing climate that had an impact on humans. Today, recurrent droughts continue to cause hardship to herders and cultivators along the southern borders of the Sahara in particular. Because the numbers of humans and animals are now far greater than they were a few thousand years ago, numbering tens of millions rather than a few thousands, their impact on the land is correspondingly far more severe. In our final chapter we look at how plants, animals, and human societies have adapted successfully to living in the arid environments of the present-day Sahara.

CHAPTER 10

Life in the Sahara:
Adapting to Aridity

The nomads will not burn the good pasture bushes . . . even in their
enemies' country . . . I have sometimes unwittingly offended them,
until I knew the plants, plucking up and giving to the flames some
which grew in the soil nigh my hand.

CHARLES M. DOUGHTY, *PASSAGES FROM ARABIA DESERTA*

Saharan Survival

There is plenty of water in the Sahara, but most of it is below the surface.
Furthermore, when there is surface water, it is often too salty to be drink-
able. The Ancient Mariner's terrible dilemma was to be surrounded by
undrinkable sea water and to see his sailing companions dying of thirst,
for which he blamed himself.[1]

As we saw in chapter 7, humans have occupied the Sahara at least
intermittently for more than half a million years, from Early Stone Age
times onwards. Whether the first humans to occupy the Sahara only did
so during wetter climatic intervals we cannot yet say. These wetter intervals were relatively short-lived in geological terms and probably only
comprised about a tenth of each full glacial-interglacial cycle.[2] The
glacial-interglacial cycles each lasted about 100,000 years.[3] However,
apart from the warm and wet interglacial phases, there were other,

shorter cycles during which the desert blossomed once more. Brief intervals when the summer monsoon was stronger would have brought wetter conditions to the southern Sahara. During times when the westerlies reached further south, notably during times when ice caps covered much of northwest Europe, conditions were wetter along the northern Sahara. However, for most of the time in which humans were in existence it seems that the Saharan climate was either arid or on the verge of becoming arid, so that plants, animals, and human communities had plenty of opportunities to adapt to growing aridity and to hone their desert survival skills.

Living in the desert is never easy. Water is generally scarce, precipitation is unreliable, and in some years it can cease altogether. On occasion heavy and often very localised downpours can cause flash floods and extreme damage to plants, soils, and human settlements. During the hot summer months, evaporation is very high, so that all organisms living in deserts need to be able to minimise water losses from their tissues if they are to survive. During winter, temperatures can fall below freezing at night, not only at high elevations in the uplands but also out on the surrounding plains. I was taken by surprise early in our time at Adrar Bous in January 1970 when I found that the water which I had left overnight in my enamel mug had turned to ice by dawn; later that day the temperature reached nearly 90°F/30°C. Winds can be strong in the Sahara, and apart from generating sand and dust storms, they add to discomfort through the 'wind-chill' factor and can cause death from hypothermia.

In spite of the very real challenges faced by desert dwellers, they have devised a formidable battery of survival strategies which have long allowed them to flourish in these harsh natural environments. I am not simply alluding to the human communities that live today in the Sahara and along its semi-arid borderlands. Adapting to aridity and living with success in a region with such unreliable rainfall applies equally to the plants and animals that live in the Sahara.[4] The animals include not only the mammals such as the alert large-eared fennecs or desert foxes and the agile little jerboas or 'desert rats', but the entire animal catalogue of birds, fishes, reptiles, insects, amphibians, crustaceans, and arachnids.

Some of these creatures are well known and generally quite easy to spot, like the large and disconcertingly fast *solifugid* spiders or sun spiders. Others, like the scorpions and vipers, are more elusive. Treated with proper respect, they will cause no harm. I still have vivid memories from just before the wet season in semi-arid central Sudan of carefully shaking out my desert boots early each morning to remove the scorpions that had sought refuge there during the night. Many other creatures are present and often unseen. All have acquired clever ways to live and flourish under arid conditions. In this chapter we consider some of the ways in which plants, animals, and human societies have adapted to life in the arid Sahara, beginning with that single most essential requirement: water.

Finding Water

An acute knowledge of where and when to find water is essential for desert survival. Gazelle can sometimes be seen just before sunrise licking the dew off the surface of dark rocks in otherwise waterless parts of the Sahara and Arabia.[5] Enigmatic piles of dark rocks scattered across the deserts of Arabia and the Sahara seem to have been designed to trap water that condensed during the cold desert nights,[6] adding moisture to the underlying soil and enabling shrubs to germinate and provide a supply of future browse. Close observation of the noisy flocks of desert-dwelling sand grouse (family *Pteroclidae*) will reveal where they fly early each morning to find water for their young which they bring back in their specially adapted feathers. Wild animals will dig for water in the sandy beds of dry stream channels or in sheltered locations and narrow crevices between the bedrock. Early in our stay at Adrar Bous, described in the prologue, we explored the eastern margins of the Aïr Mountains and observed patches of sand where jackals dug for water and found one suitable spot. The water which trickled from this underground source was enough to sustain our small team of archaeologists at Adrar Bous for twelve weeks with no ill effects.

Once we had completed our archaeological excavations and my geological surveys at Adrar Bous, which I have described in chapter 3, archaeologist Andy Smith and I decided that it would be fun to explore the Aïr Mountains in search of rock paintings and rock engravings. We had worked every day without a break for twelve weeks at Adrar Bous and felt we had earned a relief from our former daily routine. Accordingly, we set forth by camel from Iferouane (map 1) in the heart of the mountains with our local Tuareg guide, Mamunta ben Tchoko, in search of rock art in the remoter parts of the northern Aïr Mountains. Mamunta knew the mountains well. One day he led us to an underground water hole or *guelta*. This guelta was several metres deep and as big as a modest suburban swimming pool. Engraved on the wall of the guelta was a beautiful carving of a gerenuk, a type of antelope no longer found in this region today. The water marks visible on the rock face showed that the pool had been deeper in the past. One interesting and almost biblical experience was when we moved into territory that belonged to another Tuareg clan. The matriarch who owned the well we wished to use to water our camels insisted that Mamunta recite his full genealogy before granting permission. This he did with good grace and she then regaled us with tales of her travels in younger days as far afield as the desert of southwest Libya in search of high-quality dates.

Knowledge of hidden water holes is jealously guarded by local desert nomads but can be shared in hard times. Desert dwellers often maintain flexible social networks with neighbouring groups so that when bad times come, they can move into the lands of their neighbours and make use of their natural resources until the good times return and they can move back to their ancestral lands. These reciprocal social networks are essential to survival and are maintained through trade and marriage.

Just as knowledge of where to find hidden water can be a very local matter, so, too, can knowledge of which plants can best be used to feed thirsty herds of animals and which can help improve water quality. At the height of the Sahel drought in late 1974 and early 1975, geologist Mike Talbot and I were working in the arid Wadi Azaouak basin in central Niger (map 1). We were faced with the unappealing prospect of obtaining our drinking water from the muddy pools that were all that remained of a once flowing river. I generously offered the use of my shirt to filter the green sludge, but our aristocratic young Tuareg guide curled his lips in scorn and despatched his younger brother with a small axe to obtain some pieces of bark from a particular tree. He returned with the bark. When added to a calabash of muddy water, within a couple of minutes the bark caused all the organic material to flocculate and sink to the bottom, leaving us to decant the now crystal-clear water above the sludge into our goatskin leather bags. The tree was *Boscia senegalensis*. The leaves, fruit, and wood of this tree have many well-known uses, all of which were dutifully recorded in the first edition of *The Useful Plants of West Tropical Africa*, but that particular use was not included.[7] It is now. The chemicals in the bark have much the same effect as alum, which is used as a commercial water purifying agent.

Digging shallow wells in the dry bed of desert streams and former lakes is widely practised across the Sahara and indeed in deserts worldwide. When Lake Lyadu (chapter 7, fig. 7.1) in the southern Afar Desert of Ethiopia had dried out in 1975 during the time of the Sahel drought, the local Afar women filled their goatskin water bags and watered their camels from shallow pools dug into the lake floor (plate 10.1). Each leather bag held roughly 4 gallons (15 litres) of water.

Up in the arid Red Sea Hills at Erkowit in eastern Sudan, the Hadendowa or Beja pastoral nomads also obtain their water from shallow wells dug to a depth of about 1.5 metres in the bed of ephemeral stream channels. During winter, sea fogs from the Red Sea move up into the Red Sea Hills, and Erkowit (map 1) has long been renowned as a mist oasis. That may sound romantic but the reality of trying to sleep on a concrete floor beneath a dripping ceiling in the local police post on Christmas Eve 1973 was far from romantic, both for me and the two very homesick young askaris manning this isolated post.

It was while examining the fossil shells embedded in the silty clays within one of the Hadendowa wells at Erkowit that I realised there might be some truth to the story recorded by the Greek historian and traveller Diodorus Siculus ('the Sicilian') some 2000 years ago.[8] Diodorus mentioned rather casually that cattle raiders from the Red Sea Hills used to descend on the plains, steal cattle, and vanish into impenetrable swamps up in their highland lairs. He also mentioned their use of the berries of the *Ziziphus* tree to make an intoxicating drink, presumably to celebrate a successful raid. There are no swamps in the arid Red Sea Hills today, but there were at that time, as I discovered after the snail shell samples that I collected proved to be permanent water snails that lived in perennial wetlands and had radiocarbon ages between about 2000 and 1500 years ago.[9]

Digging shallow wells to find water is nothing new. Bir Sahara and Bir Tarfawi (map 3) are two remote spots in the far south of the Western Desert of Egypt. Today this region is a barren wilderness, with no recorded rainfall. The distinguished Egyptian geologist Bahay Issawi[10] was moved to describe this area in these terms: 'The area is one of the most utterly desolate parts of the earth. No vegetation, no traces of any kind of life, absolutely nothing but vast endless sheets of sands and the sun. Only at Bir Tarfawi and Bir Sahara does seepage from underground water sustain clusters of trees. At Tarfawi, a corridor of date palms, dom palms and tamarisk trees extends 15 km [9 miles] . . . in discontinuous strips. At Bir Sahara, clumps of tamarisks are often growing on a sand dune surface for a distance of eight kilometres' [5 miles].

But it was not always thus. During the last interglacial about 125,000 years ago, the climate was wet enough for a series of lakes to exist in this now desolate area (fig. 10.1).

These lakes attracted Middle Stone Age people who lived by hunting wild game and gathering wild plant foods, all of which were abundant in what was then a savanna woodland environment. When the lakes dried out during periodic dry spells, the people dug shallow wells to obtain water. These wells are still evident to the archaeologist's trowel.[11] We see this same practice at widely separated sites across the Sahara during drier intervals.

So far, I have been talking about shallow wells that are easily dug in soft sediments in dry stream channels or the dry beds of former lakes. But many of the wells that I have seen in different parts of the Sahara are remarkably deep and have been dug using very simple tools through hard rocks like the Nubian Sandstone, which covers such a huge area in the Sahara and extends across into Arabia. Indeed, the Nubian Sandstone aquifer is the largest groundwater aquifer in the world. It covers an area of two million square kilometres and has an estimated storage

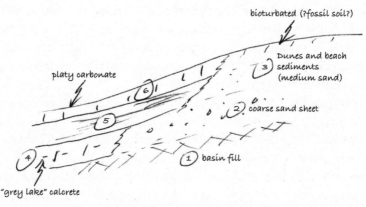

bioturbated (?fossil soil?)

platy carbonate

3) Dunes and beach sediments (medium sand)

6

2) coarse sand sheet

5

4

1) basin fill

"grey lake" calcrete

Last Interglacial lake deposits,
Bir Tarfawi, SW Egyptian Desert (18/2/87)

FIGURE 10.1. Sketch of a simplified cross-section showing two generations of lake sediments ('green lake' and 'grey lake') banked against older dune and sandy beach sediments which in turn rest on a much older basin fill deposit (18 February 1987). The lakes are about 125,000 years old and date to the Last Interglacial climatic phase, which was wet across the Sahara. The location is at Bir Tarfawi prehistoric site in the now hyper-arid far south of the Western Desert of Egypt.

of 150,000 cubic kilometres of groundwater. While the impetus to dig such deep wells came from necessity, the motivation to persevere often came from spiritual leaders.

Two centuries ago, the Zwaya tribe of Bedouin Arabs moved south into the desert from their former grazing lands in coastal Cyrenaica (northern Libya) to the remote oasis of Kufra (map 3) at the behest of their spiritual leader, the great Sufi sheikh Sayyid Muhammad ibn Ali as-Senussi (1787–1859). They learned the art of desert travel from the Tibu inhabitants of Kufra and opened up trade routes to Chad and Borkou. In the 1960s, the sixteen-day camel route from Kufra to Wanyanga (Ounianga) in Chad (map 3) was made possible by four deep wells. According to Rosita Forbes,[12] who visited Kufra with the Egyptian explorer-diplomat Ahmed Hassanein Bey during the winter of 1920–21, the sites of two of these wells were reputedly chosen by the Grand Senussi as a result of two dreams. One well took eight years to dig using fire, water, and straw bags to haul out the broken rock. Digging the other well took longer.

In these days of high-tech drilling rigs, it is easy to underestimate the ingenuity and courage of the early well diggers and their very sound practical knowledge of hydrogeology. This was brought home to me during a visit to the walled city of Harar in southeast Ethiopia with Desmond Clark in 1975. Away from the city we came across a group of men who told us they were digging a well for water, and they asked our advice as to the best place to dig. The local geology suggested to me that around the hillside would be a good spot, because permeable sandstone was sitting on granite rock with a weathered impermeable surface, and the dip of the sandstone indicated where the permeable rock would be most shallow. The men gave conspiratorial smiles and led us around the hillside to where they had already dug a well and were about to strike water. Their implements were simple stone-weighted digging sticks with iron ferrules at the tip, and straw baskets to remove the debris. I was both chastened and exhilarated.

Living with Uncertainty

Living in the Sahara means living with uncertainty. Water and food are the twin essentials for life, and some form of shelter from the elements is an advantage, particularly for the very young and the elderly. Mobile shelters can consist of tents made with drapes woven from wool obtained from sheep, goats, or camels, or more simply of a framework of pliable sticks draped in straw mats. More permanent dwellings are often made of mud bricks, or stones shaped from whatever suitable local rocks are available. In parts of Tunisia where the deposits of wind-blown dust are metres thick, underground homes are often excavated into these soft but cohesive sediments. On rare occasions, natural rock shelters or caves can provide shelter, particularly in mountainous areas.

Because rainfall is sparse and erratic, spare and intelligent use of water is vital for humans and other animals alike. Excess loss of water causes death, so that over the years desert plants and animals have developed a remarkable set of adaptive strategies, both physiological and behavioural, which we will look at in the next section. In this section I will focus on the clever ways in which the people of the desert find and use sporadic and uncertain supplies of water for themselves, their herds, and their crops, even in places where the mean annual rainfall is notionally as low as 125 millimetres (5 inches) annually. I say 'notionally' because any average figure for rainfall in the Sahara conceals more than it reveals. An average of, say, five inches/125 mm a year can mean five years with no rain at all followed by five years with about ten inches/250 mm a year. When I visited the weather station in Kufra oasis in the heart of the Libyan Desert in the northern summer of 1962, their records showed no rain for the previous twenty years, which fully explained why the main road through the oasis at that time was made of rock salt. However, an extreme downpour in 1941 caused catastrophic loss of life for many of the inhabitants of Kufra who had taken refuge in low-lying caves to escape wartime bombing, and who drowned in the ensuing flash floods. On one occasion in September 1963, while we were climbing a small sandstone hill south of Kufra, a few dark clouds began to gather during the late afternoon. With shade temperatures approaching 125°F (50°C), we were all hoping for cooling rain. We saw the rain falling, but it evaporated as it fell and never reached the ground.

Saharan rainfall is highly variable in both space and time. One locality can be deluged with rain while other places only twenty miles (thirty kilometres) away remain bone dry. Flash floods can be very dangerous in the plains surrounding the Saharan uplands. Incidents involving drowned legionaries who had camped in sandy stream channels draining the Atlas Mountains (map 1) stimulated the French military brass to draft new standing orders making it a military offence to bivouac in desert wadis, however safe and sheltered they might appear. That was over fifty years ago, but the reality of flash flooding has not changed.

Apart from the water and grazing that is available in sheltered mountain valleys for the Fur farmers of Jebel Marra, the Tibu of Tibesti and Ennedi, and the Tuareg of the Hoggar and the Aïr (map 1), there is another source of grazing known in northern Sudan and southern Libya as *gizzu* grazing, which makes its sudden appearance after local heavy rainstorms. In his account of the grasses of Africa, the Scottish botanist

J.M. Rattray[13] describes this most welcome of desert phenomena: 'In the extreme west of the Sudan there is a rather peculiar area of "gizzu" grazing. This makes its appearance if rain has fallen, with the onset of intensely cold nights. Its value to camels and sheep is great, as herds and flocks will journey 300 to 400 miles (500 to 650 kilometres) to take advantage of it. In a year without "gizzu" the animals are in much poorer condition than in a normal year when rain showers are sufficient. There is no water in the "gizzu" area and animals live entirely on the moisture derived from the grazing'. Rattray then lists eight of the more important 'gizzu' plants,[14] commenting that *Neurada procumbens* is the main provider of water for the animals.

Before motorised transport became widely available to cross the Sahara, the camel was the most reliable means of transport and is superbly adapted to desert life. Camels are vastly more useful than simply functioning as 'ships of the desert'. Camel's milk is highly nutritious and is part of the staple diet of Saharan nomads from Mauritania in the west to the Red Sea Hills in the east (map 1). When the grazing is good, the milk can be used to make cheese but is usually consumed as a liquid, often when slightly sour. Camel hair is used to make rugs and blankets. Camel hides are also used to make shelters, and great herds of camels travel from Sudan to Egypt each year to provide meat to the local markets along the Nile.

Harvesting Runoff and Early Irrigation

For 2000 years or more, desert farmers have devised some ingenious ways to harvest the runoff from flash floods and divert it to where it can best replenish the soil moisture supply while minimising possible soil losses caused by such runoff. Meticulously maintained stone terraces are common in the drier parts of Tigray province in northern Ethiopia (map 2) and in September 2019, the healthy ripening heads of barley and wheat in the small fields between terrace walls showed every sign of a bumper crop. On the volcanic slopes of Jebel Marra (map 1) in semi-arid Darfur Province in northwest Sudan, the Fur cultivators have long used the stone-banked terraces to conserve soil and water and grow their crops of sorghum, fruit, and vegetables. Some of the boulders on these terraces are very large and would have required considerable effort to move and install. The terrace walls were obviously built to last. It is not known when they were first built. In the Negev Highlands of southern Israel, various forms of terracing date back to Nabatean times some 2000 years ago, but the halcyon days of terracing were between about 300 and 1100 years ago, during Byzantine and early Islamic times.[15] They were finally abandoned because the governments of the day exacted excessive taxes from the farmers, and the prices obtained for their olives and grapes no longer justified the labour of maintaining and repairing the stone walls breached by flash floods. When properly maintained, the terrace structures enable present-day Bedouin farmers to grow viable crops of wheat and barley in this same area.

It is tempting to conclude that terracing hillsides in arid areas to conserve soil and water was a recent invention of desert farmers in North Africa and the Levant, but this may not be entirely true. Terracing hillslopes to retain soil and trap moisture from runoff is not simply the prerogative of desert farmers but has been practised by hunter-gatherers for an unknown length of time in arid northwest Australia. On rocky quartzite hillslopes in the Dampier Peninsula of the semi-arid Pilbara region of northwest Australia, the desert Aborigines built slightly curved walls of quartzite boulders to trap moisture, soil, and wind-blown dust immediately upslope of the retainer walls. They were able to harvest better crops of wild yams and other wild vegetable foods than would otherwise have been possible in this region where no rain falls for about two-thirds of the year. Similar early experiments in and around the Sahara may have taken place before the inception of cultivation and may only have developed on a large scale when the population had increased enough to provide the necessary labour.

Another very clever water harvesting practice came into use as a result of the Persian invasions of Egypt. The Persians had already developed the use of *qanats* to obtain underground water at their capital Persepolis some 2500 years ago.[16] They in turn probably learned the art from the early inhabitants of Al Ain in Oman, who were using this same system, here known as *falaj*, to harvest underground water some

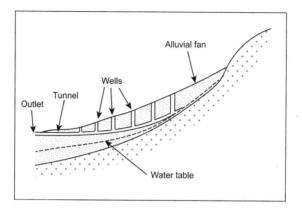

FIGURE 10.2. Simplified cross-section through a *qanat* or *foggara*.

3000 years ago, much as is done today. A qanat or falaj consists of a series of vertical shafts dug into alluvial fan sediments along the line of
steepest slope to link up with a gently sloping underground tunnel dug
so that it just intersects the top of the local water table (fig. 10.2). Water
flowing from the tunnel exit is carefully diverted to where it can irrigate
crops and fruit trees. The Persians brought the qanat concept to the
Levant and to Egypt, whence it spread across North Africa, where these
amazing underground water harvesting structures are known in Arabic
as *foggara* and their use strictly governed by traditional laws. The Arab
invaders of northwest Africa and Spain brought this clever technology
to Spain a thousand years ago and the Spanish invaders of central America brought it to Mexico about five hundred years later.

Adaptations to Aridity

During the long interval between about 15,000 and 5000 years ago the
Sahara was intermittently much less arid than it is today. I have used the
phrase 'Green Sahara' as a convenient label for this latest relatively wet
interval, although some of my friends and colleagues prefer to describe
the most recent humid episode in the Sahara as 'pale green'.[17] We have
already seen that throughout its seven million years of history as a desert,

the Sahara has experienced a large number of less arid climatic episodes, each of which could just as appropriately be termed 'when the Sahara was green'. What is important is that after each wetter climatic interval, the climate became dry once more. Plants and animals living in the Sahara would all have had to adapt to the heightened aridity, as did any early hominins and later groups of hunter-gatherers or herders and cultivators. Migration to wetter places was always an option for creatures with legs or wings. For plants such as trees, shrubs, and grasses, tethered as they were to one spot because of their root systems, migration was a much slower process, based on scattering seeds by wind and water over a wide area, with success in germinating being something of a lottery. The survival imperative demanded that plants minimise water loss and make best use of sporadic rain through a combination of physiological and behavioural adaptations to aridity. Their key requirement is to keep or acquire enough water for effective photosynthesis.

Succulent plants such as the spectacular *Euphorbia candelabra* trees that grace the summits of the Red Sea Hills retain water in their leaves, roots, and stems. Other succulents like the woody shrub *Atriplex* ('old man saltbush') with its hundreds of species can store water and tolerate salty soil, as in the gypsum-rich salt plains north of the Aurès Mountains in Algeria (map 1). The Tamarix tree is another salt-tolerant plant (or *halophyte*) and has the ability to excrete salt from its leaves. Many succulents that grow along the shores of desert salt lakes and salt pans (or *sebkhas*) share this adaptive capacity.

Succulents are however quite rare in the Sahara, where most of the plants are *xerophytes* or another group of plants that are well adapted to drought. This adaptation takes different forms. At its most extreme, the plant can survive desiccation and from being apparently dead can spring to life when the rare rains return. Lichens, which in fact consist of two organisms, a fungus and an alga, are a prime example. Xerophytic algae play an important role in forming crusts and stabilising dune surfaces.[18] Another mechanism for coping with lack of rain is to remain inactive during the dry season and only resume growth when the rains return. Those plants that do remain active during the dry season reduce water loss from transpiration by shedding bark, leaves, and even branches.

Many Saharan trees and shrubs have spines or needles in lieu of large leaves. In some trees, the needles are jointed and can be shed in segments to avoid water loss. The roots of desert plants can be wide and shallow to profit from light rain, and/or deep enough to tap into the local water table. Plant ecologists like Imanuel Noy-Meir (1941–2009) who worked all their lives in deserts have identified a number of ways in which plant communities have adapted to erratic and sometimes heavy rain, including the elegant 'pulse-response' model and the more elaborate 'pulse-reserve-response' model.[19] In essence, desert plants make opportunistic use of whatever water comes their way by responding only to certain high rainfall events, which promote growth and allow the plant to establish sufficient reserves to tide them over through the drier intervals.

Animals living in the Sahara need to minimise water loss and avoid heat stress by day and cold by night. The woollen coats of goats, sheep, and camels provide insulation against both heat and cold. Desert mammals can drink large amounts of water when necessary. They excrete highly concentrated solids and liquids with little loss of water. Resting by day and hunting by night avoids water loss and heat stress. Scorpions, lizards, and beetles stay in their cool and relatively humid burrows during the day and forage at night. Desert snakes such as the beautiful saw-scaled viper (*Echis carinatus*) can burrow into the sand and emerge when they need to warm up and increase their metabolic rate.

A further adaptation to living in the Sahara was long-distance trade. Trade in salt[20] has long allowed Saharan desert dwellers to exchange salt for other vital necessities as well as cherished luxury items such as coffee, tea, and spices. The great North African Berber scholar, geographer, and traveller Muhammad Ibn Ibrahim Ibn Battutah[21] (1304–1369) described the trade in salt, gold, and slaves in some detail in the book describing his travels. On his return to Tangier in Morocco after nearly thirty years of travel to Mecca, Persia, Arabia, Mesopotamia, Asia Minor, Bokhara, Afghanistan, China, Spain, and Timbuktu, he dictated a detailed account of the history, geography, and customs of the places he had visited, with the apt title of *Rihlah* or *Travels*. During his travels, Ibn Battutah covered more than 120,000 kilometres or 75,000 miles—a remarkable achievement given the dangers, discomforts, and difficulties of overland travel and long sea voyages at that time. There is a large sandstone plateau to the north of Tibesti in the desert of southeast Libya that now bears his name.[22]

An alternative to trading is raiding. Camels, dried dates, and goatskin water bags enabled Tuareg and Tibu raiders to travel huge distances across the Sahara in search of booty. Another great Tunisian Arab scholar-traveller and historian 'Abd Ar-Rahman Ibn Khaldun[23] (1332–1406) wrote a lengthy history of the Arabs and Berbers (*Kitab al 'Ibar*) followed by an account of the rise and fall of desert oases (the *Maqaddimah* or *Introduction to History*). He considered that once the virile and frugal desert nomads had conquered and occupied a particular desert oasis, they became seduced by easy living, lost their martial vigour, and in turn proved to be easy pickings for later, more martial invaders from the desert. This cyclical view of human history with its strong undertones of environmental determinism is reminiscent of the once popular but now outmoded views about the rise and fall of world civilisations put forward at great length by the erudite historian Arnold Toynbee.[24] The harsh life of desert nomads has always attracted the admiration of romantics such as the English writer and inveterate traveller Bruce Chatwin, but it is all too easy to forget how tough their life can be, as anyone with even a minimal knowledge of doctoring and first aid will know.

Lost Oases, Dried-Up Springs, and Water from the Rocks

The interval between the two world wars only lasted about thirty years but was a period of almost feverish exploration and mapping in the eastern Sahara by small teams of army officers from Britain and Italy in particular, who were now equipped with desert-worthy vehicles and were preparing for another possible war. They even formed an elite club of desert explorers based in Cairo whose ostensible motive was to locate some of the 'lost oases' of the Libyan Desert, including the fabled oasis of Zerzura. This oasis features briefly in Michael Ondaatje's 1992 novel

The English Patient[25] (and the film of that name) together with the enig-matic Hungarian desert explorer Count László Almásy and his 1933 dis-covery at Wadi Sura in the western Gilf Kebir (map 3) of the rock paint-ings in the sandstone rock shelter now popularly known as the 'cave of the swimmers'.[26]

After having accompanied (and greatly helped) Rosita Forbes in her journey from Benghazi to Kufra Oasis in 1921–22, the Egyptian explorer Hassanein Bey[27] made an even more adventurous journey in Decem-ber 1922 from Cairo to find the so-called lost oases (albeit well known to the Tibu and other human inhabitants of the desert) of Jebel Arkenu and Jebel 'Uweinat (map 1) in the far southeast of the Libyan Desert, where he discovered rock engravings of giraffes. He concluded that the surrounding desert at one time must have been savanna woodland for giraffes to have lived there. He was lucky to find fresh water in a small spring high up the main wadi at Jebel Arkenu. By the time I visited Jebel Arkenu in August 1962, forty years later, his former spring had dried up and had become a most unappealing mixture of rock salt and gazelle dung.[28] However, there was and still is plenty of fresh water in the four springs that emerge near the base of Jebel 'Uweinat, which has long served as an occasional refuge for the Tibu during times of hardship in their customary mountain homes in Ennedi and Tibesti (map 1). The Tibu[29] are an ancient people who have occupied the drier regions of Chad, Niger, Libya, and Sudan since well before the Arab migrations into North Africa more than a thousand years ago. Today they practice nomadic pastoralism and some cultivation.

Ralph Bagnold of the Royal Engineers was one of the leading British desert explorers and has left vivid and highly readable accounts of his many journeys taken in converted Model-T and Model-A Fords. Bag-nold's expeditions travelled overland from Cairo across the Western Desert of Egypt and through the sand seas in the southeast Libyan Des-ert as far as Jebel 'Uweinat with its freshwater springs. On one occasion Bagnold and his companions made an uncomfortable trip from 'Uweinat through the desert scrub of northern Sudan as far as the town of El Fasher (map 1) in the far northwest of Darfur Province. On their return journey they were in effect retracing the old slave route known as the *Darb al Arba'in*. What emerged very clearly from Bagnold's repeated journeys into the Libyan Desert was an abundance of evidence, in the forms of rock art, prehistoric stone tools, and fossil remains of fish, freshwater snails, and large savanna herbivores, that the hyper-arid eastern Sahara had once been capable of supporting abundant life.

In his idiosyncratic, entertaining, and informative 1990 book *Water Shall Flow from the Rock: Hydrogeology and Climate in the Lands of the Bible*, Arie Issar[30] makes a good case for the ability of those who dwell in arid lands to find water. In particular, he notes that the Bedouin Arabs of the Sinai (map 1) know where and how to excavate the softer portions in otherwise hard rocks to find water, and he points out that crossing the Sinai Desert would have been possible even for a large group of people fleeing Egyptian oppression more than 3000 years ago because reliable sources of permanent water were available within an easy day's walking distance, just as they still are today.

Summing Up

It is misleading to assume that once the Sahara became arid, life was no longer possible there. It is true that life in the Sahara is not easy and has required an intelligent and skilful series of adaptations by the human communities that call the desert home. Likewise, the plants and animals that live in the Sahara have long adapted to living with aridity. There are lessons we can learn from this remarkable success story that can help us cope with future change elsewhere. We need to possess, or to develop, a very clear-sighted understanding of the land we live in, such as the subtle web of ecosystem processes and the likely environmental hazards such as drought, fire, flood, and cyclone. Flexibility and adaptability will be essential for our future well-being on a planet that is becoming increasingly crowded and increasingly damaged by our actions to a far greater extent than ever occurred in the past. Above all, we need to learn to think long-term, and to cherish the wisdom of past peoples who learned to live in greater harmony with their land than many of us have presently managed. Finally, we need to educate our politicians to think outside the box and put aside their present obsession with the next electoral cycle and encourage our business leaders to forego short-term profit in favour of social justice and sustainable ecosystems.

In the epilogue that follows this chapter, it seems appropriate to ask whether the Sahara could become green once more and what we as humans can do to live in harmony with our greatest desert as well as with the drier regions of the earth more generally. Failure to do so will be short-sighted and self-destructive; courage in grasping this nettle will be of lasting benefit, both in terms of fostering biodiversity and in terms of social and economic rewards. How sad it is that so many of our so-called political leaders have such a myopic view of the world we live in and seem to think that ill-informed rhetoric and destructive environmental policies are an adequate substitute for what is best described as ecologically sustainable development.

Will the Sahara Become
Green Once More?

The Sahara is like a vast and extraordinary natural museum. Because it is so dry, a lot of the evidence of its past is remarkably well preserved. On the many occasions I have travelled in or across the Sahara, I have always come away deeply impressed by the evidence that the Sahara has been very different in the past, and well able to support an abundance of plants and animals as well as widely scattered groups of prehistoric hunters and gatherers and, more recently, herders and farmers. The plants, animals, and pastoralists have long since vanished except in a very few favoured localities where water is still available all year round. The questions that my Saharan journeys encouraged me to ask were generally the same. When and why did the Sahara become a desert? Why was it able to support more life in the past? Could human activities have caused the demise of what was once a 'green and pleasant land'? If so, could the Sahara become green once more? I think that we are now able to offer reasonably solid answers to each of these questions, which is why I wrote this book.

The birth of the Sahara as a desert extends back to about seven million years ago. A number of different factors contributed to the onset of aridity at this time. None had anything to do with human impact. The most important factor was the slow northward migration of Africa into latitudes characterised by dry subsiding air. A second and almost equally important factor was the sheer size of the Sahara. From the Atlantic coast to the Red Sea, the Sahara extends nearly 5000 kilometres from west to east and is on average about 2000 kilometres in breadth from south to north. As a consequence, rain-bearing winds have lost most of their moisture by the time they reach the Sahara. Only the highest Saharan mountains receive any significant precipitation today. Global cooling seven million years ago also contributed to Saharan aridity, in particular by reducing evaporation from the surrounding tropical oceans, thereby reducing tropical rainfall. The shrinking of the great Tethys Sea, of which the present-day Mediterranean Sea is but a shrunken remnant, also meant that less moisture was available along the northern Sahara.

The Sahara was not always dry during the last seven million years. Indeed, there were often long intervals when the climate was wet enough to allow big rivers to flow across what is now desert and large lakes to support an aquatic fauna of fishes, crocodiles, turtles, and hippos. The last time that the Sahara was significantly wetter was between about 15,000 and 5000 years ago. This was a time when much of the Sahara was covered in tropical savanna woodland and grassland. Lakes large and small were common everywhere. Some were fed from groundwater, which was much closer to the surface at that time. Others were fed by a combination of rainfall and local runoff. Savanna herbivores roamed the Sahara, their presence recorded in many thousands of rock engravings and rock paintings at suitable localities across the Sahara. In this rock art we see depicted elephants, giraffes, rhinos, antelopes, crocodiles, hippos, and other animals common today in the savanna country of East Africa. Somewhat later we see paintings of people with herds of cattle. The men depicted have spears and bows and arrows and are often accompanied by dogs in scenes showing wild sheep and antelopes being hunted. This long interval during which life in the Sahara was abundant is popularly termed the time when the Sahara was green. It was not to last. From about 5000 years on the climate became increasingly more arid, forcing most of the savanna fauna and prehistoric pastoralists to migrate away from the expanding desert in search of reliable sources of water and pasture.

Just as the birth of the Sahara seven million years ago was entirely independent of any human activity, so too were the alternating intervals of wetter and drier climate. Fluctuations in the amount of solar radiation received in the tropics controlled the summer rainfall regime and particularly the strength of the summer monsoon, which governed how far north into the Sahara the summer rains could penetrate. During times in the last few million years when high northern latitudes received less solar radiation than today, enabling large ice caps to build up and persist over North America and Europe, the Westerlies rainfall belt was shunted further south, bringing occasional winter rain to the northern Sahara. The reasons for these fluctuations in the amount of solar radiation received at different times in different latitudes are linked to long-term fluctuations in the tilt of the earth's axis, with a periodicity of about 40,000 years, and changes in the shape of the earth's orbit around the sun, which is sometimes more elliptical and sometimes less so, with a periodicity of about 100,000 years. The third factor is related to the season of the year in which the earth is closest to the sun and is controlled by the direction in which the spin axis of the earth points in space, with an average cycle duration of about 20,000 years. Hence, the long-term fluctuations in Saharan climate are determined by astronomical factors over which we have no control. The answer to the question will the Sahara become green once more is yes, but not for a long time to come.

In the introduction to this book, I also indicated that we need to ask what humans can do to live in harmony not only with our greatest desert but also with the drier regions of the earth more generally. Distinguished Cambridge University geographer Professor Bill Adams has worked on these issues in Africa south of the Sahara for over forty years and his insights are based on sustained first-hand practical experience. The third edition of his book, *Green Development: Environment and Sustainability in a Developing World*, discusses the intricate web of social, economic, and policy links between sustainable development and issues of dryland degradation, biodiversity conservation, and water use.[1]

However, the most succinct and, for me, intellectually compelling set of guidelines that I have so far come across are those proposed by the Swedish polymath and cancer specialist, Dr Karl-Henrik Robèrt, in his 1992 book *Det Nödvändiga Steget* (Swedish for *The Natural Step*).[2] He based his ideas on Einstein's proposition, summarised in his special theory of relativity, that matter and energy are interchangeable. I will cut to the chase and simply mention the most basic preconditions that Robèrt considered necessary for achieving any form of sustainable land use, whether in our drylands or in other regions of the world.

The first and most obvious but most overlooked point to remember is that on planet earth, the only source of a *net* increase in primary productivity is from solar energy acting through photosynthesis to increase plant matter. Everything else is simply applying energy to recycling earth materials (such as coal and iron) into other forms of matter (such as steel and carbon dioxide). It follows that for photosynthesis to operate effectively, we must protect and ideally increase the plant biomass on this earth and not allow it to be wantonly destroyed.

The second prerequisite for achieving what I think is best described as ecologically sustainable development is that we should not *systematically* remove materials from any natural or humanly modified system at a rate faster than that at which they can be replaced. A good example here is the need to ensure that the soil we all rely upon is maintained or replenished at about the same rate as that at which it will tend to form as a result of rock or sediment weathering and biological activity. David Montgomery[3] has shown us what happens when we do not nurture our soils in his eloquently argued 2007 book, *Dirt: The Erosion of Civilizations*.

The third condition is almost a mirror image of the second. It is that we should not *systematically* add materials to a natural or humanly modified system at a rate faster than the capacity of that system to absorb and recycle those materials. Chemical pollution of our air, water, and soil is a good example of what happens when we breach this principle. Over-enthusiastic use of pesticides, herbicides, and fertilizers can often do more harm than good, especially in the long-term. I recall chatting with a Tunisian subsistence farmer who had decided to adopt minimum tillage when growing wheat in a drought-prone area rather than use herbicides or fire to kill weeds. He commented that although his yields were lower in any given year, his plants survived the drought when those of other farmers in the area had perished. He concluded that it made better economic sense to adopt a long-term perspective.

A fourth and final principle is one based on simple human justice. This one is perhaps the hardest to achieve. It states that all people of the earth should have fair and effective access to natural resources. Reliable and easy access to clean and safe drinking water is a case in point.

These guidelines may seem unduly utopian and too hard to put into practice, but I do think that our survey of environmental changes in the Sahara covered in the previous ten chapters offers some grounds for cautious optimism. In spite of repeated and sometimes quite rapid changes in Saharan climates in prehistoric and historic times, many plants and animals have successfully adapted to these changes and have developed efficient and ingenious ways to cope with the harsh extremes of living in our largest desert. So, too, have those human groups who chose to remain in the Sahara even during times of prolonged aridity, when survival depended on their accumulated knowledge of where and how to find water and food for themselves and their herds. We are an adaptable species and can learn from our past. Sound policies need good science, but good science does not on its own guarantee good policies. The trick is to educate our decision makers and to ensure that they understand and accept that we need to live in harmony with all the denizens of our planet.

NOTES

Chapter 1

1. Williams, M.A.J. and Hall, D.N. (1965). Recent expeditions to Libya from the Royal Military Academy, Sandhurst. *Geographical Journal, 131*, 482–501. Describes the geology and landforms of Jebel Arkenu in southeast Libya.

2. Rognon, P. (1967). *Le Massif de l'Atakor et ses bordures (Sahara centrale). Étude géomorphologique.* CNRS, Paris, 559 pp. Gives a detailed account of the Hoggar and its environmental history.

3. Vail, J.R. (1976). Outline of the geochronology and tectonic units of the Basement Complex of north-east Africa. *Proceedings of the Royal Society of London, 350A*, 127–141. Provides a comprehensive discussion of Basement Complex rocks.

4. Williams, M. (2019). *The Nile Basin: Quaternary Geology, Geomorphology and Prehistoric Environments.* Cambridge University Press, Cambridge and New York, 405 pp. Chapter 4 provides a non-technical summary of these events.

5. Williams, F.M. (2016). *Understanding Ethiopia: Geology and Scenery.* Springer, Dordrecht, 343 pp. Gives a lucid, nontechnical account of the formation and disintegration of Gondwana. For more detail, see also Kearey, P. and Vine, F.J. (1996). *Global Tectonics*, 2nd edition. Blackwell, Oxford, 333 pp.

6. Williams, M. (2014). *Climate Change in Deserts: Past, Present and Future.* Cambridge University Press, Cambridge and New York, 629 pp. See chapter 18, as well as Williams (2019), chapter 4.

7. Vail, J.R. (1978). Outline of the geology and mineral deposits of the Democratic Republic of the Sudan and adjacent areas. *Overseas Geology and Mineral Resources* No. 49, 68 pp. Her Majesty's Stationary Office, London. See also Williams (2019), chapter 13.

8. Moorhouse, G. (1974). *The Fearful Void.* Hodder & Stoughton, London, 288 pp. He describes in graphic terms his walk from Mauritania to the central Sahara.

9. Benanav, M. (2006). *Men of Salt: Crossing the Sahara on the Caravan of White Gold.* Lyons Press, Guilford, CT, 256 pp. A vivid account of the age-old Saharan trade in salt.

10. Drake, N.A., Blench, R.M., Armitage, S.J., Bristow, C.S., and White, K.H. (2011). Ancient watercourses and biogeography of the Sahara explain the peopling of the desert. *Proceedings of the National Academy of Sciences, 108*, 458–462.

11. Drake, N. and Bristow, C. (2006). Shorelines in the Sahara: Geomorphological evidence for an enhanced monsoon from palaeolake Megachad. *The Holocene, 16*, 901–911 provides details. See also the pioneering studies by Grove, A.T. and Pullan, R.A. (1963). Some aspects of the

Pleistocene paleogeography of the Chad Basin. In *African Ecology and Human Evolution*, F. Clark Howell and F. Bourlière (eds.). Aldine, Chicago, pp. 230–245; and by Grove, A.T. and Warren, A. (1968). Quaternary landforms and climate on the south side of the Sahara. *Geographical Journal*, *134*, 194–208.

12. Armitage, S.J., Drake, N.A., Stokes, S., El-Hawat, A., Salem, M.J., White, K., Turner, P., and McLaren, S.J. (2007). Multiple phases of North African humidity recorded in lacustrine sediments from the Fazzan Basin, Libyan Sahara. *Quaternary Geochronology*, *2*, 181–186.

13. The Nubian Sandstone contains the largest store of freshwater of any rock formation and has long been vital to life in the Sahara through its wells and springs.

14. Prasad, G., Lejal-Nicol, A., and Vaudois-Meija, N. (1986). A Tertiary age for Upper Nubian Sandstone Formation, Central Sudan. *American Association of Petroleum Geologists Bulletin*, *70*, 138–142. On the basis of plant fossils, Gisela Prasad and her colleagues propose a Cenozoic age for some of the Nubian Sandstone in central Sudan.

Chapter 2

1. For a recent but fairly technical account of this event see: Fritz, H., Abdelsalam, M., Ali, K.A., Bingen, B., Collins, A.S., Fowler, A.R., Ghebreab, W., Hauzenberger, C.A., Johnson, P.R., Kusky, T.M., Macey, P., Muhongo, S., Stern, R.J., and Viola, G. (2013). Orogen styles in the East African Orogen: A review of the Neoproterozoic to Cambrian tectonic evolution. *Journal of African Earth Sciences*, *86*, 65–106.

2. For a clear illustration and explanation, see Macdougall, J.D. (1996). *A Short History of Planet Earth*. Wiley, Chichester, 266 pp. For more details, see Kearey, P. and Vine, F.J. (1996). *Global Tectonics*, 2nd edition. Blackwell, Oxford, 333 pp.

3. Williams (2014), pp. 364–368 summarises the climatic impacts of these events: Williams, M. (2014). *Climate Change in Deserts: Past, Present and Future*. Cambridge University Press, Cambridge and New York, 629 pp.

4. Black, R. and Girod, M. (1970). Late Palaeozoic to recent igneous activity in West Africa and its relationship to basement structure. In *African Magmatism and Tectonics*, T.N. Clifford and I.G. Gass (eds.). Oliver & Boyd, Edinburgh, pp. 185–210. They place the ring-complexes into a wider geological context.

5. For a pioneering study with some remarkable geological mapping by one person, see Black, R. (1963). Notes sur les complexes annulaires de Tchouni-Zarniski et de Gouré (Niger). *Bulletin de recherches géologiques et minières*, *1*, 31–45.

6. Bowden, P., Van Breemen, O., Hutchison, J., and Turner, D.C. (1976). Palaeozoic and Mesozoic age trends for some ring-complexes in Niger and Nigeria. *Nature*, *259*, 297–299. The first reliable ages for the ring-complexes in this region.

7. Francis and Oppenheimer (2004), pp. 39–43 provide a well-illustrated account of hot spots/mantle plumes and their links to volcanic activity. See Francis, P. and Oppenheimer, C. (2003). *Volcanoes*, 2nd edition. Oxford University Press, Oxford, New York, 521 pp.

8. Le Quellec (2009); Zboray (2009); and Menardi Noguera and Zboray (2011) all provide some well-illustrated descriptions. See Le Quellec, J.-L. (2009). Les images rupestres du Jebel el-'Uweynât. *Archéo-Nil*, *19*, 13–26; Zboray, A. (2009). *Rock Art of the Libyan Desert*. (DVD).

2nd edition. Newbury, Fliegel Jezernicky Expeditions Ltd; and Menardi Noguera, A. and Zboray, A. (2011). Rock art in the landscape setting of the western Jebel Uweinat (Libya). *Sahara*, 22, 85–116.

9. See Menardi Zoguera and Zboray (2011), p. 101 and plate A5, p. 111. Menardi Noguera, A. and Zboray, A. (2011). Rock art in the landscape setting of the western Jebel Uweinat (Libya). *Sahara*, 22, 85–116.

10. Professor Théodore Monod pointed this out to me in a letter some years ago.

11. See Sonntag, C., Thorweite, R.J., Lohnert, E.P., Junghans, C., Munnick, K.O., Klitzsch, E., El Shazly, E.M., and Swailem, F.M. (1980). Isotopic identification of Saharan groundwater— groundwater formation in the past. *Palaeoecology of Africa*, 12, 159–171; Joseph, A., Frangi, J.P., and Aranyossy, J.F. (1992). Isotope characteristics of meteoric water and groundwater in the Sahelo-Sudanian zone. *Journal of Geophysical Research*, 97 (No. D7), 7543–7551; Osmond, J.K. and Dabous, A.A. (2004). Timing and intensity of groundwater movement during Egyptian Sahara pluvial periods by U-series analysis of secondary U in ores and carbonates. *Quaternary Research*, 61, 85–94; and Sultan, M., Sturchio, N., Hassan, F.A., Hamdan, M.A.R., Mahmood, A.M., El Alfy, Z., and Stein, T. (1997). Precipitation source inferred from stable isotopic composition of Pleistocene groundwater and carbonate deposits in the Western Desert of Egypt. *Quaternary Research*, 48, 29–37.

12. Beuf et al. (1971) provide a thorough description of the evidence for glaciation in the northwest Sahara at this time. Beuf, S., Bijou-Duval, B., De Charpal, O., Rognon, P., et al. (1971). Les Grés du Paléozoique inférieur au Sahara. *Publication de l'Institut français du Pétrole*. Technip, Paris.

13. After a short visit to see the evidence for himself, Rhodes Fairbridge comments on how well preserved he found the glacial grooves created during this very early glaciation. Fairbridge, R.W. (1970). An ice age in the Sahara. *Geotimes*, 15, 18–20.

14. For more detail, see Macdougall (1996), pp. 168–172. See note 2 in this chapter.

15. Williams (2014), pp. 22–29 summarises the events that led up to the glaciations of the Quaternary Period. See note 3 in this chapter.

16. Messerli et al. (1980) review the evidence for recent glaciation in the Sahara and its borders. See Messerli, B., Winiger, M., and Rognon, P. (1980). The Saharan and East African uplands during the Quaternary. In *The Sahara and the Nile: Quaternary Environments and Prehistoric Occupation in Northern Africa*, M.A.J. Williams and H. Faure (eds.). A.A. Balkema, Rotterdam, pp. 87–132.

17. Grove, A.T. and Warren, A. (1968). Quaternary landforms and climate on the south side of the Sahara. *Geographical Journal*, 134, 194–208.

18. Owen, R. (1841). Report on British Reptiles, Part II. Report of the British Association for the Advancement of Science, pp. 60–204.

19. For full details, see Maidment, S.C.R., Raven, T.J., Ouarhache, D., and Barrett, P.M. (2020). North Africa's first stegosaur: Implications for Gondwanan thyreophoran dinosaur diversity. *Gondwana Research*, 77, 82–97.

20. For a thorough review of the evidence, see Russell, D.A. and Paesler, M.A. (2003). Environments of Mid-Cretaceous Saharan dinosaurs. *Cretaceous Research*, 24, 569–588.

21. Femke Holwerda gives chapter and verse in Holwerda (2020), and in her beautifully illustrated account in 'The Conversation' of 11 February 2020. See Holwerda, F.M. (2020).

Sauropod dinosaur fossils from the Kem Kem and extended 'Continental intercalaire' of North Africa. *Journal of African Earth Sciences, 163,* 103738.2.

22. Sallam, H.M. and nine others (2018). New Egyptian sauropod reveals Late Cretaceous dinosaur dispersal between Europe and Africa. *Nature Ecology & Evolution, 2,* 445–451.

23. The classic paper on dinosaur extinction caused by asteroid impact is that of Alvarez et al. (1980), written before the impact crater was discovered. See Alvarez, L.W., Alvarez, W., Asaro, F., and Michel, H.V. (1980). Extraterrestrial cause for the Cretaceous-Tertiary extinction. *Science, 208,* 1095–1108.

24. Acting under orders from the French Emperor Napoleon Bonaparte, Sonnini de Manon-court travelled widely in Egypt and returned with samples of fossil wood collected in Wadi el Natrun near the monastery dedicated to Saint Makarios. The 1807 English translation of his account (Sonnini de Manoncourt, 1807) published in London changed the French subtitle to avoid any mention of Napoleon. Sonnini de Manoncourt, C.N.S. (1807). *Travels in Upper and Lower Egypt, Undertaken by Order of the Old Government of France.* (Trans. Henry Hunter). John Stockdale, London.

25. See Williams et al. (1998), p. 117, for a more detailed explanation. Williams, M., Dunker-ley, D., De Deckker, P., Kershaw, P., and Chappell, J. (1998). *Quaternary Environments,* 2nd edi-tion. Arnold, London, 329 pp.

26. The Tethys Sea separated Gondwana and Laurasia during much of the Mesozoic Era until the breakup of Gondwana and the opening of the Indian and Atlantic Oceans during Cretaceous times.

27. Williams et al. (1987) summarise the detailed French geological research on this topic. See Williams, M.A.J., Abell, P.I., and Sparks, B.W. (1987). Quaternary landforms, sediments, depositional environments and gastropod isotope ratios at Adrar Bous, Tenere Desert of Niger, south-central Sahara. In *Desert Sediments: Ancient and Modern,* L. Frostick and I. Reid (eds.). Geological Society Special Publication No. 35, 105–125.

28. Ibid.

29. See Williams (2014), pp. 22–36, for a comprehensive overview. Williams, M. (2014). *Climate Change in Deserts: Past, Present and Future.* Cambridge University Press, Cambridge and New York, 629 pp.

30. Swezey, C.S. (2009). Cenozoic stratigraphy of the Sahara, Northern Africa. *Journal of African Earth Sciences, 53,* 89–121.

31. Avni et al. (2012) and Avni (2017) describe this major erosional event. Avni, Y. (2017). Tectonic and physiographic settings of the Levant. In *Quaternary of the Levant,* Y. Enzel and O. Bar-Yosef (eds.). Cambridge University Press, Cambridge, pp. 3–16; and Avni, Y., Segev, A., and Ginat, H. (2012). Oligocene regional denudation of the northern Afar dome: Pre- and syn-breakup stages in the Afro-Arabian plate. *Geological Society of America Bulletin, 124,* 1871–1897.

32. Griffin (2006) and Brunet et al. (2005) give details. See Griffin, D.L. (2006). The late Neogene Sahabi rivers of the Sahara and their climatic and environmental implications for the Chad Basin. *Journal of the Geological Society of London, 163,* 905–921; and Brunet, M., Guy, F., Pilbeam, D., Lieberman, D.E., Likius, A., Mackaye, H.T., Ponce de León, M., Zollikofer, C.P.E., and Vignaud, P. (2005). New material of the earliest hominid from the Upper Miocene of Chad. *Nature, 434,* 752–755.

33. Williams (1984), p. 37. Williams, M.A.J. (1984). Geology. In *Key Environments: Sahara Desert*, J.L. Cloudsley-Thompson (ed.). Pergamon Press, Oxford, pp. 31–39.

34. See Williams (2019), pp. 181–184, for a full account of these exciting finds. Williams, M. (2019). *The Nile Basin: Quaternary Geology, Geomorphology and Prehistoric Environments.* Cambridge University Press, Cambridge and New York, 405 pp.

Chapter 3

1. Williams, M. (2014). *Climate Change in Deserts: Past, Present and Future.* Cambridge University Press, Cambridge and New York, 629 pp. See p. 336 in that volume.

2. deMenocal, P., Ortiz, J., Guilderson, T., Adkins, J., Sarnthein, M., Baker, L., and Yarusinsky, M. (2000). Abrupt onset and termination of the African Humid Period: Rapid responses to gradual insolation forcing. *Quaternary Science Reviews, 19*, 347–361.

3. Drake, N.A., Blench, R.M., Armitage, S.J., Bristow, C.S., and White, K.H. (2011). Ancient watercourses and biogeography of the Sahara explain the peopling of the desert. *Proceedings of the National Academy of Sciences, 108*, 458–462.

4. Lézine, A.-M., Casanova, J., and Hillaire-Marcel, C. (1990). Across an early Holocene humid phase in western Sahara: Pollen and isotope stratigraphy. *Geology, 18*, 264–267.

5. Coulson, D. and Campbell, A. (2001). *African Rock Art: Paintings and Engravings on Stone.* H.A. Abrams, New York.

6. Williams, M.A.J. (2008). Geology, geomorphology and prehistoric environments. In *Adrar Bous: Archaeology of a Central Saharan Granitic Ring Complex in Niger*, D. Gifford-Gonzalez (ed.). Royal Museum for Central Africa, Tervuren, Belgium, pp. 25–54.

7. Drake et al. (2011). See note 3 in this chapter.

8. Clark, J.D., Carter, P.L., Gifford-Gonzalez, D., and Smith, A.B. (2008). The Adrar Bous cow and African cattle. In *Adrar Bous: Archaeology of a Central Saharan Granitic Ring Complex in Niger*, D. Gifford-Gonzalez (ed.). Royal Museum for Central Africa, Tervuren, Belgium, pp. 355–368.

9. Lhote, Henri (1959). *The Search for the Tassili Frescoes: The Story of the Prehistoric Rock Paintings of the Sahara.* Hutchinson, London.

10. Menardi Noguera, A. and Zboray, A. (2011). Rock art in the landscape setting of the western Jebel Uweinat (Libya). *Sahara, 22*, 85–116.

11. Gifford-Gonzalez, D. with Parham, J. (2008). The fauna from Adrar Bous and surrounding areas. In *Adrar Bous: Archaeology of a Central Saharan Granitic Ring Complex in Niger*, D. Gifford-Gonzalez (ed.). Royal Museum for Central Africa, Tervuren, Belgium, pp. 313–353.

12. Drake, N. and Bristow, C. (2006). Shorelines in the Sahara: Geomorphological evidence for an enhanced monsoon from palaeolake Megachad. *The Holocene, 16*, 901–911.

13. Armitage, S.J., Drake, N.A., Stokes, S., El-Hawat, A., Salem, M.J., White, K., Turner, P., and McLaren, S.J. (2007). Multiple phases of North African humidity recorded in lacustrine sediments from the Fazzan Basin, Libyan Sahara. *Quaternary Geochronology, 2*, 181–186.

14. Wendorf et al. (1993) describe many generations of former lakes dating back to more than 125,000 years ago. See Wendorf, F., Schild, R., and Close, A. (eds.) (1993). *Egypt During the Last Interglacial: The Middle Paleolithic of Bir Tarfawi and Bir Sahara East.* Plenum, New York, 596 pp.

15. Pesce, A. (1968). *Gemini Space Photographs of Libya and Tibesti: A Geological and Geographical Analysis*. Petroleum Exploration Society of Libya (Tripoli), 81 pp.

16. See Breed, C.S., McCauley, J.F., and Davis, P.A. (1987). Sand sheets of the eastern Sahara and ripple blankets on Mars. In *Desert Sediments: Ancient and Modern*, L. Frostick and I. Reid (eds.). Geological Society Special Publication No. 35, 337–359; and McHugh, W.P., Breed, C.S., Schaber, G.G., McCauley, J.F., and Szabo, B.J. (1988). Acheulian sites along the 'radar rivers', southern Egyptian Sahara. *Journal of Field Archaeology*, 15, 361–379.

17. Monod, T. (1963). The Late Tertiary and Pleistocene in the Sahara. In *African Ecology and Human Evolution*, F. Clark Howell and F. Bourlière (eds.). Aldine, Chicago, pp. 116–229.

18. See ibid. Also see Quézel, P. (1962). A propos de l'olivier de Lapérrine de l'Adrar Greboun. In *Missions Berliet Ténéré-Tchad*, H.J. Hugot (ed.). Arts et Métiers Graphiques, Paris, pp. 329–332; Wickens, G.E. (1976). *The Flora of Jebel Marra (Sudan Republic) and Its Geographical Affinities*. Kew Bulletin Additional Series V, HMSO, 368 pp.; and Maley, J. (1980). Les changements climatiques de la fin du Tertiaire en Afrique: leur conséquence sur l'apparition du Sahara et de sa végétation. In *The Sahara and the Nile*, M.A.J. Williams and H. Faure (eds.). A.A. Balkema, Rotterdam, pp. 63–86.

19. Drake et al. (2011). See note 3 in this chapter.

20. See illustrations on pp. 190–191 in Smith, A.B. (2008a). The Kiffian. In *Adrar Bous: Archaeology of a Central Saharan Granitic Ring Complex in Niger*, D. Gifford-Gonzalez (ed.). Royal Museum for Central Africa, Tervuren, Belgium, pp. 179–199.

21. Williams (2014), pp. 86–91 provides a nontechnical explanation of radiocarbon dating. See note 1 in this chapter.

22. Smith (2008a). See note 20 in this chapter.

23. See Ker Than (1 June 2011). Wormlike parasite detected in ancient mummies. *National Geographic News*; and Sandle, T. (2013). Pharaohs and mummies: Diseases of Ancient Egypt and modern approaches. *Journal of Infectious Diseases & Preventive Medicine*, 1(4), 1–2, 1000e110.

24. Gasse, F. (2000a). Water resources variability in tropical and subtropical Africa in the past. In *Water Resources Variability in Africa during the XXth Century*. International Association of Hydrological Sciences Publication No. 252, 97–105; and Gasse, F. (2000b). Hydrological changes in the African tropics since the Last Glacial Maximum. *Quaternary Science Reviews*, 19, 189–211.

25. Clark et al. (2008). See note 8 in this chapter.

26. Arkell, A.J. (1949). *Early Khartoum*. Oxford University Press, London, 145 pp.; and Arkell, A.J. (1953). *Shaheinab*. Oxford University Press, London, 114 pp.

27. Smith, A.B. (2008b). The Tenerian. In *Adrar Bous: Archaeology of a Central Saharan Granitic Ring Complex in Niger*, D. Gifford-Gonzalez (ed.). Royal Museum for Central Africa, Tervuren, Belgium, pp. 201–243.

28. Honegger, M. and Williams, M. (2015). Human occupations and environmental changes in the Nile valley during the Holocene: The case of Kerma in Upper Nubia (Northern Sudan). *Quaternary Science Reviews*, 130, 141–154.

29. Williams, M.A.J. (2009). Late Pleistocene and Holocene environments in the Nile basin. *Global and Planetary Change*, 69, 1–15.

30. Sereno, P.C. and 16 others (2008). Lakeside cemeteries in the Sahara: 5000 years of Holocene population and environmental change. *PLoS ONE*, 3(8), e2995, 1–22.

31. Ibid., see table 8.

32. Alley, R.B., Mayewski, P.A., Sowers, T., Stuiver, M., Taylor, K.C., and Clark, P.U. (1997). Holocene climatic instability: A prominent, widespread event 8200 yr ago. *Geology*, 25, 483–486.

33. Clark et al. (2008). See note 8 in this chapter.

34. Brass, M. (2017). Early North African cattle domestication and its ecological setting: A reassessment. *Journal of World Prehistory*, doi.org/10.1007/s10963-017-91122-9.

35. Williams (2019), pp. 301–320. See Williams, M. (2019). *The Nile Basin: Quaternary Geology, Geomorphology and Prehistoric Environments.* Cambridge University Press, Cambridge and New York, 405 pp.

36. Bell (1971) gives a graphic account of the demise of the Old Kingdom in Egypt. See Bell, B. (1971). The Dark Ages in Ancient History. I. The First Dark Age in Egypt. *American Journal of Archaeology*, 75, 1–26.

37. See Cullen, H.M., deMenocal, P.B., Hemming, S., Hemming, G., Brown, F.H., Guilderson, T., and Sirocko, F. (2000). Climate change and the collapse of the Akkadian Empire: Evidence from the deep sea. *Geology*, 28, 379–382; and Weiss, H. (2000). Beyond the Younger Dryas: Collapse as adaptation to abrupt climate change in ancient West Asia and the Eastern Mediterranean. In *Environmental Disaster and the Archaeology of Human Response*, G. Bawden and R.M. Reycraft (eds.). Maxwell Museum of Anthropology, University of New Mexico, Albuquerque, Anthropological Papers No. 7, pp. 75–98.

38. Singh, G. (1971). The Indus valley culture seen in the context of post-glacial climatic and ecological studies in northwest India. *Archaeology and Physical Anthropology in Oceania*, 6, 177–189.

39. Herodotus, Book 4, p. 304. See Herodotus (1954, reprinted 1960). *The Histories.* Translated by Aubrey de Sélincourt. Penguin, Middlesex.

40. The Hyksos invaded Egypt from Asia Minor in about 1660 BC (Joleaud, 1931; Williams and Hall, 1965, p. 495). See Joleaud, L. (1936). Essai stratigraphique sur les faunes des mammiferes quaternaires et leur relation avec les hommes fossils du Sahara. *Proceedings of the XVI International Geological Congress, 1933.* Washington, 1936, 789–797; and Williams, M.A.J. and Hall, D.N. (1965). Recent expeditions to Libya from the Royal Military Academy, Sandhurst. *Geographical Journal*, 131, 482–501.

41. Von Däniken (1968). An entertaining work of science fiction. See Von Däniken, E. (1968). *Chariots of the Gods? Unsolved Mysteries of the Past.* Trans. Michael Heron. Putnam, New York.

42. Lhote (1959). See note 9 in this chapter.

43. Ibid.

44. See Quézel, P. (1962). A propos de l'olivier de Lapérrine de l'Adrar Greboun. In *Missions Berliet Ténéré-Tchad*, H.J. Hugot (ed.). Arts et Métiers Graphiques, Paris, pp. 329–332; Quézel, P. (1997). High mountains of the Central Sahara: Dispersal, origin and conservation of the flora. In *Reviews in Ecology, Desert Conservation and Development: A Festschrift for Prof. M. Kassas on the Occasion of His 75th Birthday*, H.N. Barakat and A.K. Hegazy (eds.). UNESCO, IDRC and South Valley University, Cairo, Egypt, pp. 159–175; Quézel, P. and Martinez, C. (1958–59). Le dernier interpluvial au Sahara central. *Libyca*, 6–6, 211–225; and Quézel, P. and Martinez, C. (1962). Premiers résultats de l'analyse palynologique de sédiments recueillis au Sahara méridional

à l'occasion de la mission Berliet-Tchad. In *Missions Berliet Ténéré-Tchad*, H.J. Hugot (ed.). Arts et Métiers Graphiques, Paris, pp. 313–327.

45. See Ritchie, J.C. and Haynes, C.V. (1987). Holocene vegetation zonation in the eastern Sahara. *Nature*, *330*, 645–647; and Ritchie, J.C., Eyles, C.H., and Haynes, C.V. (1985). Sediment and pollen evidence for an early to mid-Holocene humid period in the eastern Sudan. *Nature*, *314*, 352–355.

46. See note 44 in this chapter.

47. Dhir, R.P., Kar, A., Wadhawan, S.K., Rajaguru, S.N., Misra, V.N., Singhvi, A.K., and Sharma, S.B. (1992). *Thar Desert in Rajasthan: Land, Man and Environment*. Geological Society of India, Bangalore, 191 pp.

48. Coulson and Campbell (2001). See note 5 in this chapter.

49. Bagnold (1935), p. 166. See Bagnold, R.A. (1935). *Libyan Sands: Travel in a Dead World*. Immel, London (1987), 288 pp.

50. Linstädter, J. and Kröpelin, S. (2004). Wadi Bakht revisited: Holocene climate change and prehistoric occupation in the Gilf Kebir region of the Eastern Sahara, SW Egypt. *Geoarchaeology*, *19*, 753–778.

51. Marinova, M.M., Meckler, A.N., and McKay, C.P. (2014). Holocene freshwater carbonate structures in the hyper-arid Gebel Uweinat region of the Sahara Desert (Southwestern Egypt). *Journal of African Earth Sciences*, *89*, 50–55.

52. Borda (2011), p. 127. See Borda, M. (2011). New painted shelter at Jebel Arkenu (Libya). *Sahara*, *22*, 125–129.

53. See Sutton, J.E.G. (1974). The aquatic civilization of middle Africa. *Journal of African History*, *15*, 527–546; and Sutton, J.E.G. (1977). The African Aqualithic. *Antiquity*, *51*, 25–34.

Chapter 4

1. Ehrlich and Ehrlich (1970), p. 202. See Ehrlich, P.R. and Ehrlich, A.H. (1970). *Population, Resources, Environment: Issues in Human Ecology*. W.H. Freeman, San Francisco, 383 pp.

2. Williams (2014), p. 308. See Williams, M. (2014). *Climate Change in Deserts: Past, Present and Future*. Cambridge University Press, Cambridge and New York, 629 pp.

3. Webster, P.J. (2004). The coupled monsoon system. In *The Asian Monsoon*, B. Wang (ed.). Springer, Berlin, pp. 3–66.

4. Williams (2014), p. 331. See note 2 in this chapter.

5. Maley (1980); Williams (2019), p. 39. See Maley, J. (1980). Les changements climatiques de la fin du Tertiaire en Afrique: leur conséquence sur l'apparition du Sahara et de sa végétation. In *The Sahara and the Nile*, M.A.J. Williams and H. Faure (eds.). A.A. Balkema, Rotterdam, pp. 63–86; and Williams, M. (2019). *The Nile Basin: Quaternary Geology, Geomorphology and Prehistoric Environments*. Cambridge University Press, New York, 405 pp.

6. Williams (2019), p. 82. See note 5 above.

7. Talbot, M.R. and Williams, M.A.J. (2009). Cenozoic evolution of the Nile basin. In *The Nile. Monographiae Biologicae, 89*, H.J. Dumont (ed.). Springer, Dordrecht, pp. 37–60.

8. Schuster, M., Duringer, P., Ghienne, J.-F., Vignaud, P., Mackaye, H.T., Likius, A., and Brunet, M. (2006). The age of the Sahara desert. *Science*, *311*, 821.

9. Habicht, J.K.A. (1979). *Paleoclimate, paleomagnetism, and continental drift.* AAPG Studies in Geology No. 9. American Association of Petroleum Geologists, Tulsa, Oklahoma.

10. Holmes (1965), p. 1054, fig. 763. See Holmes, A. (1965). *Principles of Physical Geology*, 2nd edition. Nelson, London, 1288 pp.; and Burke, K. and Gunnell, Y. (2008). The African erosion surface: A continental-scale synthesis of geomorphology, tectonics, and environmental change over the past 180 million years. *Geological Society of America Memoir, 201,* 1–66.

11. See Rognon, P. and Williams, M.A.J. (1977). Late Quaternary climatic changes in Australia and North Africa: A preliminary interpretation. *Palaeogeography, Palaeoclimatology, Palaeoecology, 21,* 285–327; and Flohn, H. (1980). The role of the elevated heat source of the Tibetan Highlands for the large-scale atmospheric circulation (with some remarks on paleoclimatic changes). *Proceedings of Symposium on Qinghai-Xizang (Tibet) Plateau (abstracts),* Beijing, China, May 25–June 1, 1980. Academia Sinica, Beijing.

12. Quade, J., Cerling, T.E., and Bowman, J.R. (1989). Development of the Asian monsoon revealed by marked ecological shift during the latest Miocene in northern Pakistan. *Nature, 342,* 163–166.

13. Sarnthein, M., Thiede, J., Pflaumann, U., Erlenkeuser, H., Fütterer, D., Koopmann, B., Lange, H., and Seibold, E. (1982). Atmospheric and oceanic circulation patterns off Northwest Africa during the past 25 million years. In *Geology of the Northwest African Continental Margin*, U. von Rad, K. Hinz, M. Sarnthein, and E. Seibold (eds.). Springer-Verlag, Berlin, pp. 545–604. See also Williams (2014), p. 336, and note 2 in this chapter.

14. Muhs, D.R., Meco, J., Budahn, J.R., Skipp, G.L., Betancort, J.F., and Lomoschitz, A. (2019). The antiquity of the Sahara Desert: New evidence from the mineralogy and geochemistry of Pliocene paleosols on the Canary Islands, Spain. *Palaeogeography, Palaeoclimatology, Palaeoecology, 533,* 109–245.

15. See Mercer, J.H. (1978). Glacial development and temperature trends in the Antarctic and in South America. In *Antarctic Glacial History and World Palaeoenvironments*, E.M. van Zinderen Bakker (ed.). A.A. Balkema, Rotterdam, pp. 73–93; and Van Zinderen Bakker, E.M. (1978). Late Mesozoic and Tertiary palaeoenvironments of the Sahara region. In *Antarctic Glacial History and World Palaeoenvironments*, E.M. van Zinderen Bakker (ed.). A.A. Balkema, Rotterdam, pp. 129–135.

16. Galloway, R.W. (1965). A note on world precipitation during the last glaciation. *Eiszeitalter und Gegenwart, 16,* 76–77.

17. Williams (2014), pp. 335–336. See note 2 in this chapter.

18. Williams, M.A.J. (2009). Cenozoic climates in deserts. In *Geomorphology of Desert Environments*, 2nd edition. A.J. Parsons and A.D. Abrahams (eds.). Springer, Berlin and New York, pp. 799–824.

19. Greigert, J. and Pougnet, R. (1967). *Essai de description des formations géologiques de la République du Niger.* Mémoires du Bureau de Recherches Géologiques et Minières (Dakar) No. 48, 1–236.

20. Williams, M.A.J., Abell, P.I., and Sparks, B.W. (1987). Quaternary landforms, sediments, depositional environments and gastropod isotope ratios at Adrar Bous, Tenere Desert of Niger, south-central Sahara. In *Desert Sediments: Ancient and Modern*, L. Frostick and I. Reid (eds.). Geological Society Special Publication No. 35, pp. 105–125.

21. Zhang, Z., Ramstein, G., Schuster, M., Li, C., Contoux, C., and Yan, Q. (2014). Aridification of the Sahara Desert caused by Tethys Sea shrinkage during the Late Miocene. *Nature*, 513, 401–404.

22. Herbert, T.D., Lawrence, K.T., Tzanova, A., Peterson, L.C., Caballero-Gill, R., and Kelly, C.S. (2016). Late Miocene global cooling and the rise of modern ecosystems. *Nature Geoscience*, 9(11), 843–849.

23. Williams et al. (1987). See note 20 in this chapter; Williams (2009). See note 18 in this chapter; Williams (2014), p. 333. See note 2 in this chapter.

24. Sarnthein et al. (1982). See note 13 in this chapter; Williams (2014), p. 336. See note 2 in this chapter.

25. See Wickens, G.E. (1976). *The Flora of Jebel Marra (Sudan Republic) and Its Geographical Affinities*. Kew Bulletin Additional Series V, HMSO, London, 368 pp.; Maley (1980). See note 5 in this chapter; and Quézel, P. (1997). High mountains of the Central Sahara: Dispersal, origin and conservation of the flora. In *Reviews in Ecology, Desert Conservation and Development: A Festschrift for Prof. M. Kassas on the Occasion of His 75th Birthday*. H.N. Barakat and A.K. Hegazy (eds.). UNESCO, IDRC and South Valley University, Cairo, Egypt, pp. 159–175.

26. See Griffin, D.L. (1999). The late Miocene climate of northeastern Africa: Unravelling the signals in the sedimentary succession. *Journal of the Geological Society of London*, 156, 817–826; Griffin, D.L. (2002). Aridity and humidity: Two aspects of the late Miocene climate of North Africa and the Mediterranean. *Palaeogeography, Palaeoclimatology, Palaeoecology*, 182, 65–91; Griffin, D.L. (2006). The late Neogene Sahabi rivers of the Sahara and their climatic and environmental implications for the Chad Basin. *Journal of the Geological Society of London*, 163, 905–921; and Griffin, D.L. (2011). The late Neogene Sahabi rivers of the Sahara and the hamadas of the eastern Libya-Chad border area. *Palaeogeography, Palaeoclimatology, Palaeoecology*, 309, 176–185.

27. Williams (2019), p. 228, fig. 16.1. See note 5 in this chapter.

28. Pesce (1968), fig. 16. See Pesce, A. (1968). *Gemini Space Photographs of Libya and Tibesti: A Geological and Geographical Analysis*. Petroleum Exploration Society of Libya (Tripoli), 81 pp.

29. Moussa, A. and 18 others (2016). Lake Chad sedimentation and environments during the late Miocene and Pliocene: New evidence from mineralogy and chemistry of the Bol core sediments. *Journal of African Earth Sciences*, 118, 192–204.

30. See Hsü, K.J. (1983). *The Mediterranean Was a Desert. A Voyage of the Glomar Challenger*. Princeton University Press, Princeton, NJ; Hsü, K.J., Montadert, L., Bernoulli, D., Cita, M.B., Erikson, A., Garrison, R.E., Kidd, R.B., Mélières, F., Müller, C., and Wright, R. (1977). History of the Mediterranean salinity crisis. *Nature*, 267, 399–403; and Williams (2019), p. 41. See note 5 in this chapter.

31. Talbot and Williams (2009), pp. 39–40. See note 7 in this chapter; Williams (2019), p. 293. See note 5 in this chapter.

Chapter 5

1. Bagnold, R.A. (1935). *Libyan Sands: Travel in a Dead World*. Immel, London (1987), 288 pp.; Bagnold, R.A. (1941). *The Physics of Blown Sand and Desert Dunes*. Methuen, London; and

Bagnold, R.A. (1990). *Sand, Wind, and War: Memoirs of a Desert Explorer.* University of Arizona Press, Tucson, 209 pp.

2. Griffin, D.L. (1999). The late Miocene climate of northeastern Africa: Unravelling the signals in the sedimentary succession. *Journal of the Geological Society of London, 156,* 817–826; Griffin, D.L. (2002). Aridity and humidity: Two aspects of the late Miocene climate of North Africa and the Mediterranean. *Palaeogeography, Palaeoclimatology, Palaeoecology, 182,* 65–91; Griffin, D.L. (2006). The late Neogene Sahabi rivers of the Sahara and their climatic and environmental implications for the Chad Basin. *Journal of the Geological Society of London, 163,* 905–921; and Griffin, D.L. (2011). The late Neogene Sahabi rivers of the Sahara and the hamadas of the eastern Libya-Chad border area. *Palaeogeography, Palaeoclimatology, Palaeoecology, 309,* 176–185.

3. See Beucher, F. (1971). *Étude palynologique de formations néogenes et quaternaries au Sahara nord-occidental,* 3 vols., Centre National de la Recherche Scientifique, Paris, 796 pp.; and Mabbutt, J.A. (1977). *Desert Landforms.* Australian National University Press, Canberra.

4. See Osborne, A.H., Vance, D., Rohling, E.J., Barton, N., Rogerson, M., and Fello, N. (2008). A humid corridor across the Sahara for the migration of early modern humans out of Africa 120,000 years ago. *Proceedings of the National Academy of Sciences, 105,* 16444–16447; Drake, N.A., Blench, R.M., Armitage, S.J., Bristow, C.S., and White, K.H. (2011). Ancient watercourses and biogeography of the Sahara explain the peopling of the desert. *Proceedings of the National Academy of Sciences, 108,* 458–462; and Coulthard, T.J., Ramirez, J.A., Barton, N., Rogerson, M., and Brücher, T. (2013). Were rivers flowing across the Sahara during the last interglacial? Implications for human migration through Africa. *PloS ONE, 8*(9): e74834.doi:10.1371/journal.pone.0074834

5. Pachur, H.-J. and Altmann, N. (2006). *Die Ostsahara im Spätquartär: Ökosystemwandel im grössten hyperariden Raum der Erde.* Springer, Berlin, 662 pp.; and Drake et al. (2011). See note 4 in this chapter.

6. Muhs, D.R. (2004). Mineralogical maturity in dunefields of North America, Africa, and Australia. *Geomorphology, 59,* 247–269.

7. Bagnold (1940). See note 1 in this chapter; Pye, K. and Tsoar, H. (1990). *Aeolian Sand and Sand Dunes.* Unwin Hyman, London; Cooke, R., Warren, A., and Goudie, A. (1993). *Desert Geomorphology.* UCL Press, London, 526 pp.; Warren, A. (2013). *Dunes: Dynamics, Morphology, History.* Royal Geographical Society with IBG, Wiley-Blackwell, Oxford, 219 pp.; and Williams (2014), pp. 112–120. See Williams, M. (2014). *Climate Change in Deserts: Past, Present and Future.* Cambridge University Press, Cambridge and New York, 629 pp.

8. Mainguet, M. and Chemin, M.-C. (1984). Les dunes pyramidales du Grand Erg Oriental. *Travaux de l'Institut de Géographie de Reims, 59–60,* 49–60.

9. Mainguet, M., Canon, L., and Chemin, M.C. (1980). Le Sahara: géomorphologie et paléogéomorphologie éoliennes. In *The Sahara and the Nile,* M.A.J. Williams and H. Faure (eds.). A.A. Balkema, Rotterdam, pp. 17–35.

10. See McCauley, J.F., Schaber, G.G., Breed, C.S., Grolier, M.J., Haynes, C.V., Issawi, B., Elachi, C., and Blom, R. (1982). Subsurface valleys and geoarchaeology of the eastern Sahara revealed by Shuttle Radar. *Science, 218,* 1004–1020; McCauley, J.F., Schaber, G.G., McHugh, W.P., Issawi, B., Haynes, C.V., Grolier, M.J., and El Kilani, A. (1986). Paleodrainages of the eastern

Sahara—the radar rivers revisited (SIR-A/B implications for a mid-Tertiary trans-African drainage system). *Institute of Electrical and Electronics Engineers Special Volume* GE-24, 624–648; Breed, C.S., McCauley, J.F., and Davis, P.A. (1987). Sand sheets of the eastern Sahara and ripple blankets on Mars. In *Desert Sediments: Ancient and Modern*, L. Frostick and I. Reid (eds.). Geological Society Special Publication No. 35, 337–359; McHugh, W.P., Breed, C.S., Schaber, G.G., McCauley, J.F., and Szabo, B.J. (1988). Acheulian sites along the 'radar rivers', southern Egyptian Sahara. *Journal of Field Archaeology*, 15, 361–379; and McHugh, W.P., Schaber, G.G., Breed, C.S., and McCauley, J.F. (1989). Neolithic adaptation and the Holocene functioning of Tertiary palaeodrainages in southern Egypt and northern Sudan. *Antiquity*, 63, 320–336.

11. See Monod, T. (1963). The Late Tertiary and Pleistocene in the Sahara. In *African Ecology and Human Evolution*, F. Clark Howell and F. Bourlière (eds.). Aldine, Chicago, pp. 116–229; and Williams (2014), p. 284. See note 7 in this chapter.

12. See Lambert, M.R.K. (1984). Amphibians and reptiles. In *Sahara Desert*, J. L. Cloudsley-Thompson (ed.). Oxford, Pergamon, pp. 205–227; and Williams (2014), p. 284. See note 7 in this chapter.

13. See Arkell, A.J. (1964). *Wanyanga and an Archaeological Reconnaisance of the South-West Libyan Desert: The British Ennedi Expedition. 1957.* Oxford University Press, London; and Williams (2019), pp. 237–238. See Williams, M. (2019). *The Nile Basin: Quaternary Geology, Geomorphology and Prehistoric Environments.* Cambridge University Press, Cambridge and New York, 405 pp.

14. Williams, M.A.J. and Hall, D.N. (1965). Recent expeditions to Libya from the Royal Military Academy, Sandhurst. *Geographical Journal*, 131, 482–501.

15. See Sarnthein, M. (1978). Sand deserts during glacial maximum and climatic optimum. *Nature*, 272, 43–45; and Sarnthein, M., Tetzlaff, G., Koopmann, B., Wolter, K., and Pflaumann, U. (1981). Glacial and interglacial wind regimes over the eastern subtropical Atlantic and northwest Africa. *Nature*, 293, 193–196.

16. See Swezey, C. (2001). Eolian sediment response to late Quaternary climate changes: Temporal and spatial patterns in the Sahara. *Palaeogeography, Palaeoclimatology, Palaeoecology*, 167, 119–155; and Williams (2014), p. 125. See note 7 in this chapter.

17. See Roskin, J., Porat, N., Tsoar, H., Blumberg, D.G., and Zander, A.M. (2011). Age, origin and climatic controls on vegetated linear dunes in the northwestern Negev Desert (Israel). *Quaternary Science Reviews*, 30, 1649–1674; Roskin, J., Tsoar, H., Porat, N., and Blumberg, D.G. (2011). Palaeoclimate interpretations of Late Pleistocene vegetated linear dune mobilization episodes: Evidence from the northwestern Negev dunefield, Israel. *Quaternary Science Reviews*, 30, 3364–3380; and Williams (2014), p. 126. See note 7 in this chapter.

18. Stokes, S., Maxwell, T.A., Haynes, C.V., and Horrocks, J. (1998). Latest Pleistocene and Holocene sand sheet construction in the Selima Sand Sheet, Eastern Sahara. In *Quaternary Deserts and Climatic Change*, A.S. Alsharan, K.W. Glennie, G.L. Whittle, and C.G.S.C. Kendall (eds.). A.A. Balkema, Rotterdam, pp. 175–184.

19. See Ritchie, J.C., Eyles, C.H., and Haynes, C.V. (1985). Sediment and pollen evidence for an early to mid-Holocene humid period in the eastern Sudan. *Nature*, 314, 352–355; Ritchie, J.C. and Haynes, C.V. (1987). Holocene vegetation zonation in the eastern Sahara. *Nature*, 330, 645–647; and Williams (2019), pp. 223–244. See note 13 in this chapter.

20. Talbot, M.R. (1985). Major bounding surfaces in aeolian sandstones—a climatic model. *Sedimentology, 32,* 257–265.

21. Williams and Hall (1965). See note 14 in this chapter.

Chapter 6

1. Prenni, A.J., Petters, M.D., Kreidenweis, S.M., Heald, C.L., Martin. S.C., Artaxo, P., Garland, R.M., Wollny, A.G., and Pöschl, U. (2009). Relative roles of biogenic emissions and Saharan dust as ice nuclei in the Amazon basin. *Nature Geosciences, 2,* 402–405.

2. Swap, R., Garstang, M., Greco, S., Talbot, R., and Kallberg, P. (1992). Saharan dust in the Amazon Basin. *Tellus, 44B,* 133–149.

3. Maley, J. (1980). Les changements climatiques de la fin du Tertiaire en Afrique: leur conséquence sur l'apparition du Sahara et de sa végétation. In *The Sahara and the Nile,* M.A.J. Williams and H. Faure (eds.). A.A. Balkema, Rotterdam, pp. 63–86.

4. Leroy, S.A.G. and Dupont, L. (1994). Development of vegetation and continental aridity in northwestern Africa during the Late Pliocene: The pollen record of ODP Site 658. *Palaeogeography, Palaeoclimatology, Palaeoecology, 109,* 295–316; and Leroy, S.A.G. and Dupont, L. M. (1997). Marine palynology of ODP Site 658 (NW Africa) and its contribution to the stratigraphy of Late Pliocene. *Geobios, 30,* 351–359.

5. Anhuf, D., Ledru, M.-P., Behling, H., Da Cruz Jr., F.W., Cordeiro, R.C., Van der Hammen, T., Karmann, I., Marengo, J.A., De Oliveira, P.E., Pessenda, L., Siffedine, A., Albuquerque, A.L., and Da Silva Dias, P.L. (2006). Paleo-environmental change in the Amazonian and African rainforest during the LGM. *Palaeogeography, Palaeoclimatology, Palaeoecology, 239,* 510–527.

6. See Pye, K. (1987). *Aeolian Dust and Dust Deposits.* Academic Press, London; and Tchakerian, V.P. (ed.) (1995). *Desert Aeolian Processes.* Chapman & Hall, London and New York.

7. Herodotus, 1960, p. 106. See Herodotus (1960). *The Histories.* Translated by Aubrey de Sélincourt. Penguin, Middlesex.

8. See Crouvi, O., Amit, R., Enzel, Y., Porat, N., and Sandler, A. (2008). Sand dunes as a major proximal dust source for late Pleistocene loess in the Negev Desert, Israel. *Quaternary Research, 70,* 275–282; Crouvi, O., Amit, R., Enzel, Y., and Gillespie, A.R. (2010). Active sand seas and the formation of desert loess. *Quaternary Science Reviews, 29,* 2087–2098; and Enzel, Y., Amit, R., Crouvi, O., and Porat, N. (2010). Abrasion-derived sediments under intensified winds at the latest Pleistocene leading edge of the advancing Sinai-Negev erg. *Quaternary Research, 74,* 121–131.

9. Kendrew (1957), p. 215. See Kendrew, W.G. (1957). *Climatology, Treated Mainly in Relation to Distribution in Time and Place.* 2nd edition. Oxford University Press, Oxford, 400 pp.

10. Pitty, A. (1968). Particle size of Saharan dust which fell in Britain in July 1968. *Nature, 220,* 364–365.

11. Franzén, L.G., Mattsson, J.O., Mårtensson, U., Nihlén, T., and Rapp, A. (1994). Yellow snow storm over the Alps and Subarctic from dust storm in Africa, March 1991. *Ambio, 23,* 233–235.

12. Dobson, M. (1781). An account of the Harmattan, a singular African wind. *Philosophical Transactions of the Royal Society of London, 71,* 46–57.

13. See Darwin, C. (1846). An account of the fine dust which often falls on vessels in the Atlantic Ocean. *Quarterly Journal of the Geological Society of London*, 2, 26–30; and Darwin, C. (1860). *Journal of Researches into the Natural History and Geology of the Countries Visited during the Voyage of H.M.S. Beagle Round the World under Captain Fitz Roy, R.N.*, 3rd edition, reprinted 2003, Folio Society, London.

14. See Morales, C. (ed.) (1979). *Saharan Dust: Mobilization, Transport, Deposition*. Wiley, Chichester; Péwé, T.L. (ed.) (1981). *Desert Dust: Origin, Characteristics, and Effect on Man*. Geological Society of America Special Paper, 186, 303 pp.; Schütz, L., Jaenicke, R., and Pietrek, H. (1981). Saharan dust transport over the North Atlantic Ocean. In *Desert Dust: Origin, Characteristics, and Effect on Man*, T.L. Péwé (ed.), 87–100; Pye (1987) (see note 6 in this chapter); and Moreno, A., Cacho, I., Canals, M., Prins, M.A., Sánchez-Goñi, M.F., Grimalt, J.O., and Weltje, G.J. (2002). Saharan dust transport and high latitude glacial climatic variability: The Alboran Sea record. *Quaternary Research*, 58, 318–328.

15. Gierlinski, G.D. and 8 others (2017). Possible hominin footprints from the late Miocene (c. 5.7 Ma) of Crete? *Proceedings of the Geologists' Association*, 128, 697–710.

16. Böhme, M. and 12 others (2017). Messinian age and savannah environment of the possible hominin *Graecopithecus* from Europe. *PLoS ONE*, 12 (5), e0177347.

17. Wendorf, F., Schild, R., and Close, A. (eds.) (1993). *Egypt During the Last Interglacial: The Middle Paleolithic of Bir Tarfawi and Bir Sahara East*. Plenum, New York, 596 pp.

18. Zaki, A.S. and Giegengack, R. (2016). Inverted topography in the southeastern part of the Western Desert of Egypt. *Journal of African Earth Sciences*, 121, 56–61; and Zaki, A.S., Pain, C.F., Edgett, K.S., and Giegengack, R. (2018). Inverted stream channels in the Western Desert of Egypt: Synergistic remote, field observations and laboratory analysis on Earth with applications to Mars. *Icarus*, 18, doi.org/10.1016/j.icarus.2018.03.001.

19. Herodotus (1960), p. 301. See note 7 in this chapter.

20. Ibid., p. 304.

21. Ibid., p. 185.

22. Kendrew (1957), p. 216. See note 9 in this chapter; Kendrew (1961), p. 67. Kendrew, W.G. (1961). *Climates of the Continents*. Oxford University Press, Oxford, 608 pp.

23. Griffiths, J.F. and Soliman, K.H. (1972). The Northern Desert (Sahara). In *World Survey of Climatology, Volume 10, Climates of Africa*, J.F. Griffiths (ed.). Elsevier, Amsterdam, pp. 75–131.

24. Williams (2019), pp. 143–163. See Williams, M. (2019). *The Nile Basin: Quaternary Geology, Geomorphology and Prehistoric Environments*. Cambridge University Press, Cambridge and New York, 405 pp.

25. Liu, Tungsheng et al. (1985). *Loess and the Environment*. China Ocean Press, Beijing, 251 pp.

26. Dobson (1781). See note 12 in this chapter.

27. Ehrenberg, C.G. (1851). On the *infusoria* and other microscopic forms in dust showers and blood rain. *American Journal of Science*, 11, 372–389.

28. See Washington, R., Todd, M.C., Lizcano, G., Tegen, I., Flamant, C., Koren, I., Ginoux, P., Engelstaedter, S., Bristow, C.S., Zender, C.S., Goudie, A.S., Warren, A., and Prospero, J.M. (2006). Links between topography, wind, deflation, lakes and dust: The case of the Bodele

Depression, Chad. *Geophysical Research Letters, 33*(9), L09401; and Warren and 8 others (2007). Dust-raising in the dustiest place on earth. *Geomorphology, 92,* 25–37.

29. A jet stream is a narrow, fast-flowing, and meandering air current.

30. Gypsum is hydrated calcium sulphate and tends to form in salty environments.

31. Prospero, J.M. and Lamb, P.J. (2003). African droughts and dust transport to the Caribbean: Climate change implications. *Science, 302,* 1024–1027.

32. See Parkin, D.W. (1974). Trade-winds during the glacial cycles. *Proceedings of the Royal Society of London, A 337,* 73–100; Parkin, D.W. and Shackleton, N. (1973). Trade-winds and temperature correlations down a deep-sea core off the Saharan coast. *Nature, 245,* 455–457; and Sarnthein, M., Tetzlaff, G., Koopmann, B., Wolter, K., and Pflaumann, U. (1981). Glacial and interglacial wind regimes over the eastern subtropical Atlantic and north-west Africa. *Nature, 293,* 193–196.

33. deMenocal, P., Ortiz, J., Guilderson, T., Adkins, J., Sarnthein, M., Baker, L., and Yarusinsky, M. (2000). Abrupt onset and termination of the African Humid Period: Rapid responses to gradual insolation forcing. *Quaternary Science Reviews, 19,* 347–361.

Chapter 7

1. Anon. (1971), pp. 108–113. See Anon. (1971). *The Epic of Gilgamesh.* Trans. N.K. Sandars. Penguin, Middlesex, 128 pp.

2. Woodward (2014), pp. 29–32. See Woodward, J. (2014). *The Ice Age: A Very Short Introduction.* Oxford University Press, Oxford, 163 pp.

3. Ibid., pp. 32–36.

4. An amusing account is that by Glyn Daniel (1964). See Daniel, G. (1964). *The Idea of Prehistory.* Penguin, Middlesex, 186 pp.

5. Oppenheimer, S. (1998). *Eden in the East: The Drowned Continent of Southeast Asia.* Weidenfeld & Nicholson, London, 560 pp.

6. Williams (1991), p. 5. See Williams, M.A.J. (1991). Evolution of the landscape. In *Monsoonal Australia: Landscape, Ecology and Man in the Northern Lowlands,* C.D. Haynes, M.G. Ridpath, and M.A.J. Williams (eds.). Balkema, Rotterdam, pp. 5–17.

7. Williams (2019), p. 198. See Williams, M. (2019). *The Nile Basin: Quaternary Geology, Geomorphology and Prehistoric Environments.* Cambridge University Press, Cambridge and New York, 405 pp.

8. Williams (1991), pp. 13–15. See note 6 in this chapter.

9. Darwin (1871), pp. 570–571. See Darwin, Charles (1871). *The Descent of Man.* Random House, New York.

10. Ibid., p. 570.

11. Ibid., p. 571.

12. Klein (1989), pp. 184–185. See Klein, R.G. (1989). *The Human Career: Human Biological and Cultural Origins.* University of Chicago Press, Chicago and London, 524 pp.

13. Claims for much younger ages are yet to be confirmed.

14. See Clark, J.D., Asfaw, B., Assefa, G., Harris, J.W.K., Kurashina, H., Walter, R.C., White, T.D., and Williams, M.A.J. (1984). Palaeoanthropological discoveries in the Middle Awash

Valley, Ethiopia. *Nature, 307*, 423–428; Clark, J.D. (1987). Transitions: *Homo erectus* and the Acheulian: the Ethiopian sites of Gadeb and the Middle Awash. *Journal of Human Evolution, 16*, 809–826; and Williams (2014), pp. 303–305. See Williams, M. (2014). *Climate Change in Deserts: Past, Present and Future.* Cambridge University Press, Cambridge and New York, 629 pp.

15. Johanson, D.C. and Edey, M.A. (1981). *Lucy: The Beginnings of Humankind.* Granada, London, 409 pp.

16. Williams (2014), p. 306. See note 14 in this chapter.

17. Plenary address by Yohannes Haile-Selassie at the Dublin INQUA Conference in July 2019.

18. White, T. (2010). Human origins. *New Scientist,* 6 November 2010, i–viii; White, T.D., Suwa, G., and Asfaw, B. (1994). *Australopithecus ramidus,* a new species of early hominid from Aramis, Ethiopia. *Nature, 371*, 306–312; White, T.D., Suwa, G., Simpson, S., and Asfaw, B. (2000). Jaws and teeth of *Australopithecus afarensis* from Maka, Middle Awash, Ethiopia. *American Journal of Physical Anthropology, 111*, 45–68; White, T.D., Ambrose, S.H., Suwa, G., and Wolde-Gabriel, G. (2010). Response to Comment on the paleoenvironment of *Australopithecus ramidus.* *Science, 328*, 1105-e; White, T.D., WoldeGabriel, G., et al. (2006). Asa Issie, Aramis and the origin of *Australopithecus.* *Nature, 440*, 883–889.

19. Haile-Selassie, Y. (2001). Late Miocene hominids from the Middle Awash, Ethiopia. *Nature, 412*, 178–181; Haile-Selassie, Y., Saylor, B.Z., Deino, A., Levin, N.E., Alene, M., and Latimer, B.M. (2012). A new hominin foot from Ethiopia shows multiple Pliocene bipedal adaptations. *Nature, 483*, 565–569.

20. Talk given by W.W. Bishop at the 1973 INQUA Conference in Christchurch, New Zealand.

21. Bonnefille, R., Potts, R., Chalié, F., Jolly, D., and Peyron, O. (2004). High-resolution vegetation and climate change associated with Pliocene *Australopithecus afarensis. Proceedings of the National Academy of Sciences, 101*, 12125–12129.

22. Barboni, D., Bonnefille, R., Alexandre, A., and Meunier, J.D. (1999). Phytoliths as paleoenvironmental indicators, West Side Middle Awash, Ethiopia. *Palaeogeography, Palaeoclimatology, Palaeoecology, 152*, 87–100.

23. Quade, J., Levin, N., Semaw, S., Stout, D., Renne, P., Rogers, M., and Simpson, S. (2004). Paleoenvironments of the earliest stone toolmakers, Gona, Ethiopia. *Bulletin of the Geological Society of America, 116*, 1529–1544.

24. F.M. Williams (2014), pp. 39–51. See Williams, F.M. (2014). *Understanding Ethiopia: Geology and Scenery.* Springer, Heidelberg, 343 pp.

25. Brunet, M., Beauvilain, A., Coppens, Y., Heintz, E., Moutaye, A.H.E., and Pilbeam, D. (1995). The first australopithecine 2,500 kilometres west of the Rift Valley (Chad). *Nature, 378*, 273–275; Brunet, M. and 37 others. (2002). A new hominid from the Upper Miocene of Chad, Central Africa. *Nature, 418*, 145–151; and Brunet, M., Guy, F., Pilbeam, D., Lieberman, D.E., Likius, A., Mackaye, H.T., Ponce de León, M., Zollikofer, C.P.E., and Vignaud, P. (2005). New material of the earliest hominid from the Upper Miocene of Chad. *Nature, 434*, 752–755.

26. Lebatard, A.E., Bourlès, D.L., Duringer, P., Jolivet, M., Braucher, R., Carcaillet, J., Schuster, M., Arnaud, M., Monié, P., Lihoreau, F., Likius, A., Mackaye, H.T., Vignaud, P., and Brunet, M. (2008). Cosmogenic nuclide dating of *Sahelanthropus tchadensis* and *Australopithecus*

bahrelghazali: Mio-Pliocene early hominids from Chad. *Proceedings of the National Academy of Sciences, 105* (9), 3226–3231.

27. Brunet, M., Beauvilain, A., Coppens, Y., Heintz, E., Moutaye, A.H.E., and Pilbeam, D. (1996). *Australopithecus bahrelghazali,* une nouvelle espèce d'hominidé ancien de la région de Koro Toro (Tchad). *Comptes Rendus, Académie des Sciences, Paris, 322,* 907–913. See also note 26 in this chapter.

28. Goodall, Jane (1976). Continuities between chimpanzee and human behaviour. In *Human Origins: Louis Leakey and the East African Evidence,* G Ll. Isaac and E.R. McCown (eds.). Benjamin, Menlo Park, CA, pp. 81–95.

29. Roche, H. (1980). *Premiers Outils Taillés d'Afrique.* Société d'Ethnographie, Paris, 114 pp.; Kimbel, W.H., Walter, R.C., Johanson, D.C., Reed, K.E., Aronson, J.L., Assefa, Z., Marean, C.W., Eck, G.G., Bobe, R., Hovers, E., Rak, Y., Vondra, C., Yemane, T., York, D., Chen, Y., Evensen, N.M., and Smith, P.E. (1996). Late Pliocene *Homo* and Oldowan tools from the Hadar formation (Kada Hadar Member), Ethiopia. *Journal of Human Evolution, 31,* 549–561; and Semaw, S., Renne, P., Harris, J.W.K., Feibel, C.S., Bernor, R.L., Fesseka, N., and Mowbray, K. (1997). 2.5-million-year-old stone tools from Gona, Ethiopia. *Nature, 385,* 333–336.

30. Semaw et al. (1997). See note 29 in this chapter.

31. See note 29 in this chapter.

32. Klein (1989), pp. 163–170. See note 12 in this chapter; Williams (2014), pp. 308–310. See note 14 in this chapter.

33. Klein (1989), pp. 199, 210–211. See note 12 in this chapter; De Heinzelin, J., Clark, J.D., Schick, K.D., and Gilbert, W.H. (eds.) (2000). *The Acheulean and the Plio-Pleistocene Deposits of the Middle Awash Valley, Ethiopia. Musée Royal de l'Afrique Central, Tervuren, Belgium, Annales-Sciences Géologiques, 104,* 235 pp.; and Williams (2014), pp. 308–309. See note 14 in this chapter.

34. Williams (2014), p. 310. See note 14 in this chapter.

35. Clark, J.D. and Kurashina, H. (1979). Hominid occupation of the east-central highlands of Ethiopia in the Plio-Pleistocene. *Nature, 282,* 33–39; and Clark (1987). See note 14 in this chapter.

36. Clark and Kurashina (1979). See note 35 in this chapter.

37. Keeley, L. (1980). *Experimental Determination of Stone Tool Use.* University of Chicago Press, Chicago; Keeley, N.H. and Toth, N. (1981). Microwear polishes on early stone tools from Koobi Fora, Kenya. *Nature, 293,* 464–465; and Schick, K.D. and Toth, N. (1995). *Making Silent Stones Speak: Human Evolution and the Dawn of Technology.* Phoenix, London.

38. Watson (2005), p. 27. See Watson, P. (2005). *Ideas: A History from Fire to Freud.* Weidenfeld and Nicholson, London; Clark, J.D., Brown, K.S., Dietler, M., Reilly, M.B., Staley, P., and Schokkenbroek, A.M. (2008). The Late Acheulian assemblages. In *Adrar Bous: Archaeology of a Central Saharan Granitic Ring Complex in Niger,* D. Gifford-Gonzalez (ed.). Royal Museum for Central Africa, Tervuren, Belgium, pp. 55–90; and Williams, M., Glasby, P., and Blackwood, J. (2008). A note on an Acheulian biface from Adrar Bous, Tenere Desert, south central Sahara, Republic of Niger. *Sahara, 19,* 85–90.

39. Clark, J.D. and Harris, J.W.K. (1985). Fire and its role in early hominid lifeways. *African Archaeological Review, 3,* 3–27.

40. Klein (1989), pp. 185–188. See note 12 in this chapter.

41. Ibid., p. 109.

42. Williams (2014), pp. 305–307. See note 14 in this chapter.

43. Klein (1989), pp. 421–422. See note 12 in this chapter.

44. See Clark, J.D. (1982). The cultures of the Middle Palaeolithic/Middle Stone Age. In *The Cambridge History of Africa, Volume I, From the Earliest Times to c. 500 BC*, J.D. Clark (ed.). Cambridge University Press, Cambridge, pp. 248–341; and Klein (1989), pp. 309–311. See note 12 in this chapter.

45. See Clark, J.D. (1980). Human populations and cultural adaptations in the Sahara and Nile during prehistoric times. In *The Sahara and the Nile: Quaternary Environments and Prehistoric Occupation in Northern Africa*, M.A.J. Williams and H. Faure (eds.). A.A. Balkema, Rotterdam, pp. 527–582; and Clark, J.D. with Schultz, D.U., Kroll, E.M., Freedman, E.E., Galloway, A., Batkin, J., Kurashina, H., and Gifford-Gonzalez, D. (2008a). The Aterian of Adrar Bous and the central Sahara. In *Adrar Bous: Archaeology of a Central Saharan Granitic Ring Complex in Niger*, D. Gifford-Gonzalez (ed.). Royal Museum for Central Africa, Tervuren, Belgium, pp. 91–162.

46. See Clark, J.D. et al. (2003). Stratigraphic, chronological and behavioural contexts of Pleistocene *Homo sapiens* from Middle Awash, Ethiopia. *Nature, 423*, 747–752; White, T.D., Asfaw, B., DeGusta, D., Gilbert, H., Richards, G.D., Suwa, G., and Clark Howell, F. (2003). Pleistocene *Homo sapiens* from Middle Awash, Ethiopia. *Nature, 423*, 742–747; Osborne, A.H., Vance, D., Rohling, E.J., Barton, N., Rogerson, M., and Fello, N. (2008). A humid corridor across the Sahara for the migration of early modern humans out of Africa 120,000 years ago. *Proceedings of the National Academy of Sciences, 105*, 16444–16447; Brown, F.H., McDougall, I., and Fleagle, J.G. (2012). Correlation of the KHS Tuff of the Kibish Formation to volcanic ash layers at other sites, and the age of early *Homo sapiens* (Omo I and Omo II). *Journal of Human Evolution, 63*, 577–585; and Richter, D. et al. (2017). The age of the hominin fossils from Jebel Irhoud, Morocco, and the origins of the Middle Stone Age. *Nature, 546*, 293–296.

47. Hershovitz, I. et al. (2018). The earliest modern humans outside Africa. *Science, 359*, 456–459.

48. Ambrose, S.H. (2001). Paleolithic technology and human evolution. *Science, 291*, 1748–1753.

49. Clark (1980). See note 45 in this chapter.

50. Williams (2019), p. 220. The wild grasses collected today along the southern Sahara include *Brachiaria, Digitaria, Cenchrus* and *Panicum*. See note 7 in this chapter.

51. Clark (1980). See note 45 in this chapter.

52. See Gopher, A., Lev-Yadun, S., and Abbo, S. (2017). Domesticating plants in the Near East. In *Quaternary of the Levant*, Y. Enzel and O. Bar-Yosef (eds.). Cambridge University Press, Cambridge, pp. 737–742; and Kislev, M.E. and Simchoni, O. (2017). Early Agriculture in the Southern Levant. In *Quaternary of the Levant*, Y. Enzel and O. Bar-Yosef (eds.). Cambridge University Press, Cambridge, pp. 733–736.

53. Winchell, F., Stevens, C.J., Murphy, C., Champion, L., and Fuller, D.Q. (2017). Evidence for sorghum domestication in fourth millennium BC eastern Sudan. *Current Anthropology*, https://doi.org/10.1086/693898.

54. See Holdaway, S.J. and Wendrich, W. (eds.) (2017). *The Desert Fayum Reinvestigated: The Early to Mid-Holocene Landscape Archaeology of the Fayum North Shore, Egypt.* UCLA Cotsen Institute of Archaeology Press, Los Angeles, *Monumenta archaeologica, 39,* 262 pp.; and Brass, M. (2017). Early North African cattle domestication and its ecological setting: A reassessment. *Journal of World Prehistory,* doi.org/10.1007/s10963-017-91122-9.

55. Brass (2017). See note 54 in this chapter.

56. Clark, J.D., Carter, P.L., Gifford-Gonzalez, D., and Smith, A.B. (2008). The Adrar Bous cow and African cattle. In *Adrar Bous: Archaeology of a Central Saharan Granitic Ring Complex in Niger,* D. Gifford-Gonzalez (ed.). Royal Museum for Central Africa, Tervuren, Belgium, pp. 355–368.

57. Williams (2019), p. 320, table 22.1. See note 7 in this chapter.

58. Ibid.

59. Clark, J. D. (1989). Shabona: An Early Khartoum settlement on the White Nile. In *Late Prehistory of the Nile Basin and the Sahara,* L. Krzyzaniak and M. Kobusiewicz (eds.). Poznan Archaeological Museum, Poznan, Poland, pp. 387–410.

60. Sereno, P.C. and 16 others (2008). Lakeside cemeteries in the Sahara: 5000 years of Holocene population and environmental change. *PLoS ONE, 3*(8), e2995, 1–22.

61. See Hewitt, G. (2000). The genetic legacy of the Quaternary ice ages. *Nature, 405,* 907–913; Williams (2014), pp. 221–228. See note 14 in this chapter; and Williams (2019), p. 332. See note 7 in this chapter.

62. Williams (2014), p. 125, fig. 8.8. See note 14 in this chapter.

63. See Faure, H. (1966). Évolution des grands lacs sahariens à l'Holocène. *Quaternaria, 8,* 167–175; Faure, H. (1969). Lacs quaternaires du Sahara. *Internationale Vereiningung für theoretische und angewandte Limnologie, 17,* 131–146; Faure, H., Manguin, E., and Nydal, R. (1963). Formations lacustres du Quaternaire supérieur du Niger oriental: diatomites et âges absolus. *Bulletin BRGM (Dakar), 3,* 41–63; and Drake, N.A., Blench, R.M., Armitage, S.J., Bristow, C.S., and White, K.H. (2011). Ancient watercourses and biogeography of the Sahara explain the peopling of the desert. *Proceedings of the National Academy of Sciences, 108,* 458–462.

64. See Roberts, N., Taieb, M., Barker, P., Damnatti, B., Icole, M., and Williamson, D. (1993). Timing of the Younger Dryas event in East Africa from lake-level changes. *Nature, 366,* 146–148; Swezey, C. (2001). Eolian sediment response to late Quaternary climate changes: Temporal and spatial patterns in the Sahara. *Palaeogeography, Palaeoclimatology, Palaeoecology, 167,* 119–155; and Gasse, F. and Roberts, C.N. (2004). Late Quaternary hydrologic changes in the arid and semi-arid belt of northern Africa. In *The Hadley Circulation: Present, Past and Future,* H.F. Diaz and R.S. Bradley (eds.). Kluwer Academic Publishers, Dordrecht, pp. 313–345.

65. See Grove, A.T. (1993). Africa's climate in the Holocene. In *The Archaeology of Africa: Food, Metals and Towns,* T. Shaw, P. Sinclair, B. Andah, and A. Okpoko (eds.). Routledge, London, pp. 32–42; and Swezey (2001). See note 64 in this chapter.

66. Grove (1993). See note 65 in this chapter; and Garcin, Y., Vincens, A., Williamson, D., Buchet, G., and Guiot, J. (2007). Abrupt resumption of the African Monsoon at the Younger Dryas-Holocene climatic transition. *Quaternary Science Reviews, 26,* 690–704.

67. See Servant, M. and Servant-Vildary, S. (1980). L'environnement quaternaire du basin du Tchad. In *The Sahara and the Nile,* M.A.J. Williams and H. Faure (eds.). A.A. Balkema,

Rotterdam, pp. 133–162; Claussen, M., Kubatzki, C., Brovkin, V., Ganopolski, A., Hoelzmann, P., and Pachur, H.-J. (1999). Simulation of an abrupt change in Saharan vegetation in the mid-Holocene. *Geophysical Research Letters, 26*, 2037–2040; and Drake, N. and Bristow, C. (2006). Shorelines in the Sahara: geomorphological evidence for an enhanced monsoon from palaeo-lake Megachad. *The Holocene, 16*, 901–911.

68. See Ritchie, J.C. and Haynes, C.V. (1987). Holocene vegetation zonation in the eastern Sahara. *Nature, 330*, 645–647; and Ritchie, J.C., Eyles, C.H., and Haynes, C.V. (1985). Sediment and pollen evidence for an early to mid-Holocene humid period in the eastern Sudan. *Nature, 314*, 352–355.

69. Williams (2019), pp. 187–195. See note 7 in this chapter.

70. Smith, J. (1949). Distribution of tree species in the Sudan in relation to rainfall and soil texture. *Sudan Government Ministry of Agriculture*, Bulletin 4.

71. Williams (2014), pp. 120–121. See note 14 in this chapter.

72. Stemler, A.B.L. (1980). Origins of plant domestication in the Sahara and Nile valley. In *The Sahara and the Nile: Quaternary Environments and Prehistoric Occupation in Northern Africa*, M.A.J. Williams and H. Faure (eds.). A.A. Balkema, Rotterdam, pp. 503–526.

73. Holdaway and Wendrich (2017). See note 54 in this chapter.

74. Sereno et al. (2008). See note 60 in this chapter.

75. Smith, A.B. (1992). *Pastoralism in Africa: Origins and Development Ecology*. Hurst & Company, London, 288 pp.

76. Williams (2016), p. 79. See Williams, M. (2016). *Nile Waters, Saharan Sands: Adventures of a Geomorphologist at Large*. Springer, Heidelberg, 210 pp.

Chapter 8

1. Williams and Balling (1996), p. 17. See Williams, M.A.J. and Balling, R.C. Jr. (1996). *Interactions of Desertification and Climate*. Arnold, London, with WMO and UNEP, 270 pp.

2. Beran and Rodier (1985), p. 1. See Beran, M.A. and Rodier, J.A. (rapporteurs) (1985). *Hydrological Aspects of Drought*. UNESCO-WMO, Geneva.

3. Baker, J.C.A. and Spracklen, D.V. (2019). Climate benefits of intact Amazon forests and the biophysical consequences of disturbance. *Frontiers in Forests & Global Change*, 30 August 2019, doc.org/10.3389//ffgc.2019.00047.

4. Nicholson, S.E. (2003). The diurnal cycle of precipitation and cloudiness over Lake Victoria and its influence on evaporation over the lake. *Bulletin of the International Decade for East African Lakes*, Summer 2003, 1–4.

5. See Lamb, P.J. (1978a). Case studies of tropical Atlantic surface circulation patterns during recent sub-Saharan weather anomalies: 1967 and 1968. *Monthly Weather Review, 106*, 482–491; and Lamb, P.J. (1978b). Large-scale tropical Atlantic surface circulation patterns associated with Subsaharan weather anomalies. *Tellus, 30*, 240–251.

6. See Markham, C.G. and McLain, D.R. (1977). Sea-surface temperature related to rain in Ceará, north-eastern Brazil. *Nature, 265*, 320–323; Lamb, P.J. and Peppler, R.A. (1991). West Africa. In *Teleconnections Linking Worldwide Climate Anomalies*, M. Glantz, R.W. Katz, and N. Nicholls (eds.). Cambridge University Press, Cambridge, pp. 121–189; and Lamb, P.J. and

Peppler, R.A. (1992). Futher case studies of tropical Atlantic surface atmospheric and oceanic patterns associated with sub-Saharan drought. *Journal of Climate, 5,* 476–488.

7. See Camberlin, P., Janicot, S., and Poccard, I. (2001). Seasonality and atmospheric dynamics of the teleconnection between African rainfall and tropical sea-surface temperature: Atlantic vs. ENSO. *International Journal of Climatology, 21,* 973–1005; and Williams and Balling (1996), p. 158, table 7.1. See note 1 in this chapter.

8. See Dorize, L. (1974). L'oscillation pluviométrique récente sur le basin du Lac Tchad et la circulation atmosphérique générale. *Revue de Géographie Physique et de Géologie Dynamique* (2), 16(4), 393–420; Dorize, L. (1976). L'oscillation climatique actuelle au Sahara. *Revue de Géographie Physique et de Géologie Dynamique* (2), 18(2–3), 217–228; Laya, D. (1975). Interviews with farmers and livestock owners in the Sahel. *African Environment, 1–2,* 49–93; Krebs, J.R. and Coe, M.J. (1985). Sahel famine: An ecological perspective. *Nature, 317,* 13; Sinclair, A.R.E. and Fryxell, J.M. (1985). The Sahel of Africa: Ecology of a disaster. *Canadian Journal of Zoology, 63,* 487; Roche, M., Rodier, J., and Sircoulon, J. (1975). Les aspects hydrologiques de la sécheresse récente en Afrique de l'Ouest. *International Union of Geodesy and Geophysics, Grenoble 1975, Symposium 16: Meteorological and Hydrological Aspects of Continental Droughts,* Paper 2 (mimeo, preprint); and Boyd, E., Cornforth, R.J., Lamb, P.J., Tarhule, A., Lélé, M.I., and Brouder, A. (2013). Building resilience to face recurring environmental crisis in African Sahel. *Nature Climate Change, 3,* 631–637.

9. Zeng, N. (2003). Drought in the Sahel. *Science, 302,* 999–1000.

10. Lélé, M.I. and Lamb, P.J. (2010). Variability of the Intertropical Front (ITF) and rainfall over the West African Sudan-Sahel zone. *Journal of Climate, 23,* 3984–4004.

11. Bell, M.A. and Lamb, P.J. (2006). Integration of weather system variability to multidecadal regional climate change: The West African Sudan-Sahel zone, 1951–98. *Journal of Climate, 19,* 5343–5365.

12. See Nicholson, S.E. (1978). Climatic variations in the Sahel and other African regions during the past five centuries. *Journal of Arid Environments, 1,* 3–24; Street-Perrott, F.A. and Perrott, R.A. (1990). Abrupt climatic fluctuations in the tropics: The influence of Atlantic Ocean circulation. *Nature, 343,* 607–611; Mo, K., Bell, G.D., and Thiaw, W.M. (2001). Impact of sea surface temperature anomalies on the Atlantic tropical storm activity and West African rainfall. *Journal of the Atmospheric Sciences, 58,* 3477–3496; and Osman, Y.Z. and Shamseldin, A.Y. (2002). Qualitative rainfall prediction models for central and southern Sudan using El Niño-Southern Oscillation and Indian Ocean sea-surface temperature indices. *International Journal of Climatology, 22,* 1861–1878.

13. Charney, J.G. (1975). Dynamics of deserts and droughts in the Sahel. *Quarterly Journal of the Royal Meteorological Society, 101,* 193–202; Charney, J.G., Stone, P.H., and Quirk, W.J. (1975). Drought in the Sahara: A biogeophysical feedback mechanism. *Science, 187,* 434–435; and Charney, J., Quirk, W.J., Chow, S., and Kornfield, J. (1977). A comparative study of the effects of albedo change on drought in semi-arid regions. *Journal of Atmospheric Sciences, 34,* 1366–1385.

14. Williams and Balling (1996), pp. 32–34. See Williams, M.A.J. and Balling, R.C. Jr. (1996). *Interactions of Desertification and Climate.* Arnold, London, with WMO and UNEP, 270 pp.

15. Walker, G.T. (1924). Correlations in seasonal variations of weather. IX: A further study of world weather. *Memoirs of the Indian Meteorological Department, 24,* 275–332.

16. See Quinn, W.H. and Neal, V.T. (1987). El Niño occurrences over the past four and a half centuries. *Journal of Geophysical Research*, 92, 14 449–14 461; Whetton, P., Allan, R., and Rutherfurd, I. (1996). Historical ENSO teleconnections in the Eastern Hemisphere: Comparison with the latest El Niño series of Quinn. *Climatic Change*, 32, 103–109; Sandweiss, D.H., Maasch, K.A., Burger, R.L., Richardson, J.B. III, Rollins, H.B., and Clement, A. (2001). Variation in Holocene El Niño frequencies: Climate records and cultural consequences in ancient Peru. *Geology*, 29, 603–606; and Cane, M.A. (2005). The evolution of El Niño, past and present. *Earth and Planetary Science Letters*, 230, 227–240.

17. Moy, C. M., Seltzer, G. O., Rodbell, D.T., and Anderson, D. M. (2002). Variability of El Niño / Southern Oscillation activity at millennial timescales during the Holocene epoch. *Nature*, 420, 162–165.

18. See Diaz, H.F. and Markgraf, V. (eds.) (1992). *El Niño: Historical and Paleoclimatic Aspects of the Southern Oscillation*. Cambridge University Press, Cambridge, UK, 476 pp.; Diaz, H.F. and Markgraf, V. (eds.) (2000). *El Niño and the Southern Oscillation: Multiscale Variability and Global and Regional Impacts*. Cambridge University Press, Cambridge, UK, 496 pp.; Allan, R., Lindesay, J., and Parker, D. (1996). *El Niño Southern Oscillation & Climatic Variability*. CSIRO Publishing, Collingwood, Victoria, Australia, 405 pp.; and Peel, M.C., McMahon, T.A., and Finlayson, B.L. (2002). Variability of annual precipitation and its relationship to the El Nino-Southern Oscillation. *Journal of Climate*, 15, 545–551.

19. Murphy, J.O. and Whetton, P.H. (1989). A re-analysis of a tree ring chronology from Java. *Proceedings of the Koninklijke Nederlandse Akademie van Wetenschappen, Series B*, 92(3), 241–257.

20. Tapper, N. (2002). Climate, climatic variability and atmospheric circulation patterns in the maritime continent region. In *Bridging Wallace's Line: The Environmental and Cultural History and Dynamics of the SE Asian-Australian Region*, P. Kershaw, B. David, N. Tapper, D. Penny, and J. Brown (eds.), *Advances in Geoecology* 34, Verlag, Catena, pp. 5–28.

21. See Whetton, P., Adamson, D., and Williams, M. (1990). Rainfall and river flow variability in Africa, Australia and East Asia linked to El Niño–Southern Oscillation events. In *Lessons for Human Survival: Nature's Record from the Quaternary. Geological Society of Australia Symposium Proceedings*, P. Bishop (ed.), 1, 71–82; Whetton, P. H. and Rutherfurd, I. (1994). Historical ENSO teleconnections in the Eastern Hemisphere. *Climatic Change*, 28, 221–253; and Whetton et al. (1996). See note 16 in this chapter.

22. Sultan, M., Sturchio, N., Hassan, F.A., Hamdan, M.A.R., Mahmood, A.M., El Alfy, Z., and Stein, T. (1997). Precipitation source inferred from stable isotopic composition of Pleistocene groundwater and carbonate deposits in the Western Desert of Egypt. *Quaternary Research*, 48, 29–37.

23. Saji, N.H., Goswami, B.N., Vinayachandran, P.N., and Yamagata, T. (1999). A dipole mode in the tropical Indian Ocean. *Nature*, 401, 360–363.

24. Flohn, H. (1987). East African rains of 1961/62 and the abrupt change of the White Nile discharge. *Palaeoecology of Africa*, 18, 3–18.

25. Stenchikov, G., Robock, A., Ramaswamy, V., Schwarzkopf, M.D., Hamilton, K., and Ramachandran, S. (2012). Arctic Oscillation response to the 1991 Pinatubo eruption: Effects of volcanic aerosols and ozone depletion. *Journal of Geophysical Research*, 107 (D24), 4803, doi: 10.1029/2002JD002090, 2002.

26. Wood, Gillen D'Arcy (2014). *Tambora: The Eruption that Changed the World*. Princeton University Press, Princeton, NJ, 293 pp. This is a beautifully written and enthralling narrative.

27. Trenberth, K.E. and Dai, A. (2007). Effect of Mount Pinatubo volcanic eruption on the hydrological cycle as an analog of geoengineering. *Geophysical Research Letters*, 34, L15702, doi:10.1029/2007GL030524, 1–5.

28. Cook, E.R., Anchukaitis, K.J., Buckley, B.M., D'Arrigo, R.D., Jacoby, G.C., and Wright, W.E. (2010). Asian monsoon failure and megadrought during the last millennium. *Science*, 328, 486–489.

29. Nicholson, S.E. (1980). Saharan climates in historic times. In *The Sahara and the Nile*, M.A.J. Williams and H. Faure (eds.). A.A. Balkema, Rotterdam, pp. 173–200.

30. See Lamb, H.H. (1970). Volcanic dust in the atmosphere; with a chronology and assessment of its meteorological significance. *Philosophical Transactions of the Royal Society of London A*, 266 (1178), 426–533; Lamb, H.H. (1972). *Climate: Present, Past and Future. Vol. 1. Fundamentals and Climate Now*. Methuen, London, 613 pp.; and Lamb, H.H. (1977). *Climate: Present, Past and Future. Vol. 2. Climatic History and the Future*. Methuen, London, 835 pp. This is a comprehensive two-volume study of global climate changes.

31. Oman, L. Robock, A., Stenchikov, G.L., and Thordarson, T. (2006). High-latitude eruptions cast shadow over the African monsoon and the flow of the Nile. *Geophysical Research Letters*, 33, L18711, 1–5.

32. Ibid.

33. Haywood, J.M., Jones, A., Bellouin, N., and Stephenson, D. (2013). Asymmetric forcing from Stratospheric aerosols impacts Sahelian rainfall. *Nature Climate Change*, 3, 660–665.

34. See Adams, J.B., Mann, M.E., and Ammann, C.M. (2003). Proxy evidence of an El Niño-like response to volcanic forcing. *Nature*, 426, 274–278; and De Silva, S. (2003). Eruptions linked to El Niño. *Nature*, 426, 239–241.

35. Whetton, P., Adamson, D., and Williams, M. (1990). Rainfall and river flow variability in Africa, Australia and East Asia linked to El Niño–Southern Oscillation events. In *Lessons for Human Survival: Nature's Record from the Quaternary. Geological Society of Australia Symposium Proceedings*, 1, P. Bishop (ed.), pp. 71–82.

36. Ibid.

Chapter 9

1. Bryson and Murray (1977), p. 109. See Bryson, Reid A. and Murray, Thomas J. (1977). *Climates of Hunger: Mankind and the World's Changing Weather*. University of Wisconsin Press, Madison, 171 pp.

2. Singhvi, A.K., Williams, M.A.J., Rajaguru, S.N., Misra, V.N., Chawla, S., Stokes, S., Chauhan, N., Francis, T., Ganjoo, R.K., and Humphreys, G.S. (2010). A ~200 ka record of climatic change and dune activity in the Thar Desert, India. *Quaternary Science Reviews*, 29, 3095–3105.

3. Herbert, T.D., Lawrence, K.T., Tzanova, A., Peterson, L.C., Caballero-Gill, R., and Kelly, C.S. (2016). Late Miocene global cooling and the rise of modern ecosystems. *Nature Geoscience*, 9(11), 843–849.

4. Ehrlich and Ehrlich, (1972), p. 202. See Ehrlich, P.R. and Ehrlich, A.H. (1972). *Population, Resources, Environment: Issues in Human Ecology.* Freeman, San Francisco, 509 pp.

5. Zhang, Z., Ramstein, G., Schuster, M., Li, C., Contoux, C., and Yan, Q. (2014). Aridification of the Sahara Desert caused by Tethys Sea shrinkage during the Late Miocene. *Nature*, *513*, 401–404.

6. See Davis, Diana K. (2007). *Resurrecting the Granary of Rome. Environmental History and French Colonial Expansion in North Africa.* Ohio University Press, Athens, 296 pp.; and Davis, Diana K. (2016). *The Arid Lands: History, Power, Knowledge.* MIT Press, Cambridge, MA, 271 pp.

7. Lamprey, H.F. (1975). *Report on the Desert Encroachment Reconnaissance in Northern Sudan, 21 Oct. to 10 Nov. 1975.* UNESCO/UNEP, Paris/Nairobi; republished in *Desertification Control Bulletin, 17*, 1–7.

8. Ibid.

9. Harrison, M.N. and Jackson, J.K. (1958). *Ecological Classification of the Vegetation of the Sudan.* Forests Department Bulletin 2 (New Series), Ministry of Agriculture, Republic of Sudan.

10. Lélé, M.I. and Lamb, P.J. (2010). Variability of the Intertropical Front (ITF) and rainfall over the West African Sudan-Sahel zone. *Journal of Climate, 23*, 3984–4004.

11. See Tucker, C.J., Dregne, H.E., and Newcomb, W.W. (1991). Expansion and contraction of the Sahara Desert from 1980 to 1990. *Science, 253*, 299–301; and Tucker, C.J. and Nicholson, S.E. (1999). Variations in the size of the Sahara Desert from 1980 to 1997. *Ambio, 28*, 587–591.

12. See Stebbing, E.P. (1935). The encroaching Sahara. *Geographical Journal, 85*, 506–524; Stebbing, E.P. (1937a). *The Forests of West Africa and the Sahara.* Chambers, Edinburgh, 245 pp.; Stebbing, E.P. (1937b). The threat of the Sahara. *Journal of the Royal African Society*, London, Extra Supplement, *36*, 3–35; and Stebbing, E.P. (1938). The man-made desert in Africa: Erosion and drought. *Journal of the Royal African Society*, London, Extra Supplement, *37*, 3–40.

13. Tacitus (98 AD). *Agricola.*

14. Davis (2007) and Davis (2016). See note 6 in this chapter.

15. Mabbutt, J.A. (1978). Desertification of Australia in its global context. *Search, 9*, 252–256.

16. Leopold (1949), p. 194. See Leopold, Aldo (1949). *A Sand County Almanac and Sketches Here and There.* Oxford University Press, Oxford, 226 pp.

17. Smith, A.B. (1992). *Pastoralism in Africa: Origins and Development Ecology.* Hurst & Company, London, 288 pp.

18. Avni, Y., Avni, G., and Porat, N. (2019). A review of the rise and fall of ancient desert runoff agriculture in the Negev Highlands—A model for the southern Levant deserts. *Journal of Arid Environments, 163*, 127–137.

19. Vita-Finzi (1969), pp. 23–24. See Vita-Finzi, C. (1969). *The Mediterranean Valleys: Geological Changes in Historical Times.* Cambridge University Press, Cambridge, 140 pp.

20. Aubréville (1949), p. 332. See Aubréville, A. (1949). *Climats, forêts et désertification de l'Afrique tropicale.* Société d'Éditions Géographiques, Maritimes et Coloniales, Paris, 352 pp.

21. Davis (2007) and Davis (2016). See note 6 in this chapter.

22. Montgomery, D.R. (2007). *Dirt: The Erosion of Civilizations.* University of California Press, Berkeley, 285 pp.

23. Williams, M.A.J. (1969). Prediction of rainsplash erosion in the seasonally wet tropics. *Nature, 222* (5195), 763–765.

24. See Cerling, T.E., Harris, J.M., MacFadden, B.J., Leakey, M.G., Quade, J., Eisenmann, V., and Ehlerlinger, J.R. (1997). Global vegetation change through the Miocene/Pliocene boundary. *Nature, 389*, 153–158; and Cerling, T.E., Wynn, J.G., Andanje, S.A., Bird, M.I., Korir, D.K., Levin, N.E., Mace, W., Macharia, A.N., Quade, J., and Remien, C.H. (2011). Woody cover and hominin environments in the past 6 million years. *Nature, 476*, 51–56.

25. UNEP (1992). Middleton, N. and Thomas, D.S.G. (eds.), *World Atlas of Desertification*, 1st edition. Arnold, London.

26. UNEP (1992). *Status of desertification and implementation of the United Nations Plan of Action to Combat Desertification*. Report of the Executive Director. United Nations Environment Programme, Nairobi.

27. Williams (2014), p. 498. See Williams, M. (2014). *Climate Change in Deserts: Past, Present and Future*. Cambridge University Press, Cambridge and New York, 629 pp.

28. Tuan, Y.F. (1966). New Mexican gullies: A critical review and some recent observations. *Annals of the Association of American Geographers, 56*, 573–597.

29. deMenocal, P., Ortiz, J., Guilderson, T., Adkins, J., Sarnthein, M., Baker, L., and Yarusinsky, M. (2000). Abrupt onset and termination of the African Humid Period: Rapid responses to gradual insolation forcing. *Quaternary Science Reviews, 19*, 347–361.

30. Gasse, F., Chalié, F., Vincens, A., Williams, M.A.J., and Williamson, D. (2008). Climatic patterns in equatorial and southern Africa from 30,000 to 10,000 years ago reconstructed from terrestrial and near-shore proxy data. *Quaternary Science Reviews, 27*, 2316–2340.

31. Holmes, J. and Hoelzmann, P. (2017). The Late Pleistocene-Holocene African Humid period as evident in lakes. *Oxford Research Encyclopedia of Climate Science*, 44 pp. doi:10.1093/acrefore/9780190228620.013.531.

32. Kröpelin, S., Verschuren, D., Lézine, A.-M., Eggermont, H., Cocquyt, C., Francus, P., Cazet, J.-P., Fagot, M., Rumes, B., Russell, J.M., Darius, F., Coley, D.J., Schuster, M., von Suchodoletz, H., and Engstrom, D.R. (2008). Climate-driven ecosystem succession in the Sahara: The past 6000 years. *Science, 320*, 765–768.

33. Williams (2014), pp. 190–194. See note 27 in this chapter.

34. Drake, N. and Bristow, C. (2006). Shorelines in the Sahara: Geomorphological evidence for an enhanced monsoon from palaeolake Megachad. *The Holocene, 16*, 901–911.

35. Wright, D.K. (2017). Humans as agents in the termination of the African Humid Period. *Frontiers in Earth Science, 5*, Article 4, 1–14.

36. Williams (2014), pp. 377–378. See note 27 in this chapter.

37. Williams, M.A.J. (1988). After the deluge: The Neolithic landscape in North Africa. In J. Bower and D. Lubell (eds.), *Prehistoric Cultures and Environments in the Late Quaternary of Africa. Cambridge Monographs in African Archaeology, 26*, BAR International Series 405, 43–60.

38. Ibid.

Chapter 10

1. Samuel Taylor Coleridge (1798), *The Rime of the Ancient Mariner*.

2. Williams (2014), pp. 32–36. See Williams, M. (2014). *Climate Change in Deserts: Past, Present and Future*. Cambridge University Press, Cambridge and New York, 629 pp.

3. Ibid., p. 33. From about 700,000 years ago onwards, each complete glacial-interglacial cycle lasted about 100,000 years; before then the cycles were shorter.

4. See Evenari, M., Shanan, L., and Tadmor, N. (1971). *The Negev. The Challenge of a Desert.* Harvard University Press, Cambridge, 345 pp.; Evenari, M., Noy-Meir, I., and Goodall, D.W. (eds.) (1985). *Hot Deserts and Arid Shrublands, A. (Ecosystems of the World, 12A).* Elsevier, Amsterdam, 365 pp.; Evenari, M., Noy-Meir, I., and Goodall, D.W. (eds.) (1986). *Hot Deserts and Arid Shrublands, B. (Ecosystems of the World, 12B).* Elsevier, Amsterdam, 451 pp.; and Williams (2014), pp. 37–56. See note 2 in this chapter.

5. Pirenne, J. (1977). La maîtrise de l'eau en Arabie du sud antique. Six types de monuments techniques. *Mémoires de l'Académie des Inscriptions et Belles-Lettres, Institut de France, Nouvelle Série* t. II, Paris, 237 pp.

6. Ibid.

7. Burkill, H.M. (1985). *The Useful Plants of West Tropical Africa, Families A-D.* 2nd edition, volume 1. Royal Botanic Gardens, Kew, 960 pp.

8. Jackson (1957), p. 58. See Jackson, J.K. (1957). Changes in the climate and vegetation of the Sudan. *Sudan Notes and Records, 38,* 47–66.

9. Mawson, R. and Williams, M.A.J. (1984). A wetter climate in eastern Sudan 2,000 years ago? *Nature, 309,* 49–51.

10. Issawi (1976), p. 20. See Issawi, B. (1976). An introduction to the physiography of the Nile Valley. In *Prehistory of the Nile Valley,* F. Wendorf and R. Schild (eds.). Academic Press, New York, pp. 3–22.

11. Wendorf et al. (1993), pp. 525–526. See Wendorf, F., Schild, R., and Close, A. (eds.) (1993). *Egypt During the Last Interglacial: The Middle Paleolithic of Bir Tarfawi and Bir Sahara East.* Plenum, New York, 596 pp.

12. Forbes, Rosita. (1921). *The Secret of the Sahara: Kufara.* Cassell, London, 350 pp.

13. See Rattray, J.M. (1960). *The Grass Cover of Africa.* FAO Agricultural Studies No. 49, 169 pp.; see p. 50. The main 'gizzu' species he cites are *Indigofera bracteolata, Neurada procumbens, Indigofera arenaria, Aristida papposa,* and *Crotelaria thebaica.*

14. Ibid.

15. Avni et al. (2019). See Avni, Y., Avni, G., and Porat, N. (2019). A review of the rise and fall of ancient desert runoff agriculture in the Negev Highlands—A model for the southern Levant deserts. *Journal of Arid Environments, 163,* 127–137.

16. Evenari, M., Shanan, L., and Tadmor, N. (1971). *The Negev. The Challenge of a Desert.* Harvard University Press, Cambridge, 345 pp.

17. Quade et al. (2018), p. 253. See Quade, J., Dente, E., Armon, M., Ben Dor, Y., Morin, E., Adam, O., and Enzel, Y. (2018). Megalakes in the Sahara? A review. *Quaternary Research, 90,* 253–275. The authors of this stimulating paper argue that because of the poor preservation of many former lake shorelines in the Sahara, some claims about the size of these lakes remain in dispute.

18. Williams (2014), p. 42. See note 2 in this chapter.

19. See Noy-Meir, I. (1973a). Desert ecostems. I. Environment and producers. *Annual Review of Ecology and Systematics, 4,* 25–52; Noy-Meir, I. (1973b). Desert ecosystems: structure and function. In *Desert Ecosystems of the World,* M. Evenari, I. Noy-Meir, and D.W. Goodall (eds.).

Elsevier, Amsterdam, pp. 92–103; Noy-Meir, I. (1974). Desert ecosytems. II. Higher trophic levels. *Annual Review of Ecology and Systematics, 5,* 195–214. See also Williams (2014), p. 44, note 2.

20. Benanav, M. (2006). *Men of Salt: Crossing the Sahara on the Caravan of White Gold.* Lyons Press, Guilford, CT, 220 pp.

21. Mackintosh-Smith (2012), xvi–xxviii. See Mackintosh-Smith, T. (2012). *The Travels of Ibn Battutah.* Folio Society, London, 353 pp.

22. Pesce (1968), p. 28 and figure 16, p. 59. Williams and Hall (1965) named the two large sandstone plateaux further north the *Hamada el Fayoud* and the *Hamada el Akdamin*; petroleum geologist Angelo Pesce very appropriately named the third most southern plateau the *Hamada Ibn Battutah.* See Pesce, A. (1968). *Gemini Space Photographs of Libya and Tibesti: A Geological and Geographical Analysis.* Petroleum Exploration Society of Libya (Tripoli), 81 pp.; and Williams, M.A.J. and Hall, D.N. (1965). Recent expeditions to Libya from the Royal Military Academy, Sandhurst. *Geographical Journal, 131,* 482–501.

23. Hourani (2009), pp. 3–6. See Hourani, A. (2009). *A History of the Arab Peoples.* Folio Society, London, 659 pp.

24. Toynbee (1934–1961), 12 volumes. The 1972 one-volume abridged and illustrated volume by Arnold Toynbee and Jane Caplan is an excellent introduction. See Toynbee, A.J. (1934–1961). *A Study of History.* 12 volumes. Oxford University Press, Oxford; and Toynbee, A. and Caplan, J. (1972). *A Study of History* (Abridged and Illustrated). One volume. Weathervane Books, New York, 576 pp.

25. Ondaatje, M. (1992). *The English Patient.* Alfred Knopf, New York, 307 pp.

26. Zboray (2003), p. 112. See Zboray, A. (2003). New rock art findings at Jebel Uweinat and the Gilf Kebir. *Sahara, 14,* 111–127.

27. See Hassanein Bey, A.M. (1924). Through Kufra to Darfur. *Geographical Journal, 64(4),* 273–291, *64(5),* 253–366; and Hassanein Bey, A.M. (1925). *The Lost Oasis.* Butterworth, London, 316 pp.

28. Williams, M.A.J. and Hall, D.N. (1965). Recent expeditions to Libya from the Royal Military Academy, Sandhurst. *Geographical Journal, 131,* 482–501.

29. Baroin, C. (2003). *Les Toubou du Sahara Central.* Vents de Sable, Paris, 171 pp.

30. Issar (1990), pp. 109–122. See Issar, Arie (1990). *Water Shall Flow from the Rock: Hydrogeology and Climate in the Lands of the Bible.* Springer-Verlag, Berlin, 213 pp.

Epilogue

1. Adams, W.M. (2009). *Green Development: Environment and Sustainability in a Developing World.* Routledge, New York, 449 pp.

2. Robèrt, K.-H. (1992). *Det Nödvändiga Steget (The Natural Step).* Ekerlids forläg, Falun, Sweden (in Swedish).

3. Montgomery, D.R. (2007). *Dirt: The Erosion of Civilizations.* University of California Press, Berkeley, 285 pp.

FURTHER READING

Ralph A. **Bagnold** (1935). *Libyan Sands: Travel in a Dead World*. Immel, London (1987), 288 pp. A vivid and classic account of early exploration by vehicle in the eastern Sahara.

Ralph A. **Bagnold** (1990). *Sand, Wind, and War: Memoirs of a Desert Explorer*. University of Arizona Press, Tucson, 209 pp. A swashbuckling account of a life devoted to exploring and understanding the Sahara and its dunes.

David **Coulson** and Alec **Campbell** (2001). *African Rock Art: Paintings and Engravings on Stone*. Abrams, New York, 256 pp. By far the best photographs and descriptions of the magnificent rock art galleries of the Sahara and the lands further south.

René **Gardi** (1967). *Sahara*. Harrap, London, 149 pp. The best introduction to the Sahara that I know, with wonderful photos of people, desert animals, and rock pictures.

John Julius **Norwich** (1968). *Sahara*. Longmans, London, 240 pp. Great photos and lively text.

Michael **Palin** (2002). *Sahara*. Weidenfeld & Nicholson, London, 256 pp. Lively and well-illustrated account of his travels by a sympathetic observer.

Martin **Williams** (2016). *Nile Waters, Saharan Sands: Adventures of a Geomorphologist at Large*. Springer, Heidelberg and New York, 210 pp. A light-hearted account of the ups and downs of Saharan fieldwork.

Martin **Williams** (2014). *Climate Change in Deserts: Past, Present and Future*. Cambridge University Press, Cambridge and New York, 629 pp. A comprehensive account of the environmental history of all the world's hot deserts, with a careful look at the quality of the evidence used to reconstruct past environments.

In addition, NASA and Google Earth provide a stunning array of colour satellite images of the Sahara and its borders which are freely available through the Internet.

INDEX OF SUBJECTS

INDEX OF PEOPLE

INDEX OF PLACES